Better Homes and Gardens®

Indoor
Gardening Made Easy

Meredith Consumer Marketing
Des Moines, Iowa

Better Homes and Gardens.

Indoor
Gardening Made Easy

MEREDITH CONSUMER MARKETING
Vice President, Consumer Marketing: Janet Donnelly
Consumer Marketing Product Director: Heather Sorensen
Consumer Marketing Product Manager: Wendy Merical
Business Director: Ron Clingman
Senior Production Manager: Al Rodruck

WATERBURY PUBLICATIONS, INC.
Contributing Editor: Karen Weir-Jimerson, Studio G, Inc.
Contributing Copy Editor: Peg Smith
Contributing Proofreader: Terri Fredrickson
Contributing Indexer: Donald Glassman

Editorial Director: Lisa Kingsley
Creative Director: Ken Carlson
Associate Editors: Tricia Bergman, Mary Williams
Associate Design Director: Doug Samuelson
Production Assistant: Mindy Samuelson

BETTER HOMES AND GARDENS MAGAZINE
Editor in Chief: Gayle Goodson Butler
Managing Editor: Gregory H. Kayko
Creative Director: Michael D. Belknap
Deputy Editor, Gardening: Eric Liskey

MEREDITH NATIONAL MEDIA GROUP
President: Tom Harty

MEREDITH CORPORATION
Chairman and Chief Executive Officer: Stephen M. Lacy

In Memoriam: E.T. Meredith III (1933–2003)

Pictured on the front cover:
top left Yellow kalanchoe adds sunny color to indoor decor.
bottom left 'Marble Queen' pothos features beautiful variegated leaves.
right Succulents, such as *Echeveria*, *Sedum*, and *Pachyveria*, are easy to grow indoors.

All of us at Meredith Consumer Marketing are dedicated to providing you with information and ideas to enhance your home. We welcome your comments and suggestions. Write to us at: Meredith Consumer Marketing, 1716 Locust St., Des Moines, IA 50309-3023.

Contents

Chapter 1

6 INDOOR GARDENING

Discover creative and practical uses for beautiful indoor plants.

8   Bring Nature Indoors
10 Garden All Year Long
12 Enhance Health with Plants

Chapter 2

14 INDOOR PLANT DECOR

Use plants to add style, life, and better health to your home.

16 Design Basics
18 Design with Plants
20 Find Your Style
22 Make Room for Plants
24 Siting Plants
26 Special Effects
28 Ideal Environment

Chapter 3

30 PLANT CARE

Learn the fundamentals of growing plants indoors successfully.

32 The Right Light
34 Watering
36 Success Factors
38 Selecting Healthy Plants
40 Choosing Containers
42 Displaying Pots
44 Potting and Repotting
46 Soil Mixes
48 Fertilizing Plants
50 Grooming Plants
52 Plant Vacation Care
54 Pest Control
56 Beating Disease

Chapter 4

58 PLANT PROPAGATION

Discover how to make new plants from division, cuttings, and seed starting.

60 Cutting Techniques
62 Plant Division
64 Plantlets and Offsets
66 Layering Techniques
68 Starting Plants from Seed

Chapter 5

70 CREATIVE PLANTING PROJECTS

Create beautiful, easy, and inspirational indoor gardens year-round.

72 Windowsill Gardens
74 Dish Gardens
76 Little Landscapes
78 Wooden Planters
80 Basket Gardens
82 Creative Containers
84 Seasonal Centerpieces
88 Topiary Towers
90 Support Systems
92 Bonsai
94 Hanging Planters
96 Forcing Bulbs
98 Forcing Hyacinths
100 Forcing Branches
102 Succulent Wreaths

Chapter 6

104 GARDEN UNDER GLASS

Plant mini tabletop gardens that sparkle under glass.

106 Moss Gardens
108 Tabletop Conservatories
110 Terrariums
112 Bottles and Jars
114 Mini Biospheres
116 Glass Bubbles
118 Cloches

Chapter 7

120 INDOOR EDIBLES

Grow a surprising array of delicious and healthful edibles.

122 Herbs
124 Plants from Produce
126 Citrus
128 Sprouts
130 Microgreens
132 Wheatgrass

Chapter 8

134 DECORATOR FAVORITES

Add floral flourish and color to your home with these stylish plants.

136 African Violet
138 Begonia
140 Bromeliad
142 Cactus
144 Coleus
146 Fern
148 Ivy
150 Jasmine
152 Orchid
154 Scented Geranium
156 Succulents

Chapter 9

158 HOUSEPLANT ENCYCLOPEDIA

From traditional varieties to new introductions, here are 47 popular and easy-care houseplants for your home.

CHAPTER ONE

Indoor Gardening

Discover creative and practical ways
to decorate your home with plants.
Surround yourself with living things.

Bring Nature Indoors

Enjoy the beauty and serenity of the great outdoors in your home with houseplants. They add life to every room in your home.

Sunshine, lush greenery, fresh flowers, aromatic soil—these elements lure people into indoor gardening. The soothing process of coaxing plants to grow in a sunny window allows indoor gardeners to be close to nature, even when winter winds rattle the glass and snow piles up outside. Plants bring nature indoors. Adding life, producing fresh air, and reducing indoor air pollution—plants make homes more livable. Watching seeds sprout and bulbs grow produces feelings of pride, joy, and astonishment that only nature inspires.

Plants showcase a multitude of fascinating talents indoors, where they can be appreciated up close in the form of multicolor leaves, twining stems, trimmable shapes, and edible parts. Weather-defying feats prove marvelous, especially when cut branches are forced to bloom indoors, and as seeds and bulbs send up shoots and blooms long before outdoor gardens can be tilled.

Indoor havens

Potted plants create fragrant, blooming havens that are every bit as pleasurable as conventional gardens. They also contribute to good health. As plants are watered, they produce humidity, a benefit to dry rooms and arid conditions. With the haven of a home, plants flourish.

Fortunately for home gardeners, plants will thrive in a guest room or a greenhouse. All they need is a smattering of sun, a dependable supply of water, reasonably benevolent temperatures, adequate nutrients, and a whiff of fresh air now and then. Generations of indoor gardeners have discovered that successful growing hinges on ingenuity coupled with an abiding love and respect for growing plants.

For centuries, people have matched their wits against the elements in order to grow living plants near them. Ancient Romans kept their dwellings brimming with fragrant and edible plants. Many Victorians, who dared to put palms in parlors and ferns on pedestals, added heated conservatories to their homes so they could grow even more plants indoors.

A world of plants

It's possible to grow a range of indoor plants—from groundcovers to vines, shrubs, and trees, and from holiday gift plants to long-lived heirlooms. Although indoor gardening trends come and go—from macramé plant hangers to the latest interest in air plants—a passion for tropical plants is evident.

Tried-and-true houseplants remain popular. What's more, new generations of houseplants with tropical origins have become easier to grow, requiring less light and little maintenance. For examples, sturdy flamingo flower and indestructible snake plant join adaptable zeezee plant in the easy-growing category. New color patterns of Chinese evergreen and dracaena have turned houseplants into home decor accessories.

Trips to grocery or building supply stores expand indoor gardening horizons with seemingly delicate orchids or other blooming treasures. Plants once hard to find are now common and affordable, and gardeners have a world of plants available via the Internet.

opposite, far left A mini tabletop garden heralds spring with pink cyclamen, yellow and purple violas, blue grape hyacinths, and bright yellow daffodils. Reindeer moss conceals soil and holds in moisture.
opposite, top right An antique terrarium is the stage for a miniature woodland garden, with Streptocarpus, ferns, and variegated ivy.
opposite, center right Long-lasting succulents need little care and make themselves at home in a large shell. **opposite, bottom right** Plants delineate transitional spaces between outdoors and indoors.

Garden All Year Long

If you love to garden outside, indoor gardening will fill the need to dig, tend, and grow when the ground is frozen and outdoor plants are at rest.

Growing plants indoors is an act of faith and hope. The perceived benefits are motivation to poke seeds and roots into containers to watch and grow whatever the season. Plants help sustain people, providing tangible rewards (in the form of beautiful flowers, luscious fruits) as well as loftier ones (such as raising spirits, relieving stress). However many or whichever types of plants are grown, the experience of gardening indoors is enriching. Regardless of climate, fresh herbs and other edibles can be grown indoors year-round, with natural or supplemental light or both.

Spend as little as a few minutes a week to easily manage a few plants indoors. Choose tough foliage plants—that can live for decades with little more than occasional watering and fertilizing—or opt for blooming plants that need more specific and attentive care. Make indoor gardening fit your lifestyle. As you master the mechanics of growing plants indoors, explore a wider variety of intriguing plants and creative arrangements. To broaden your expertise beyond easy-care houseplants, explore, learn, and grow with this book as your guide.

For veteran growers, winter is neither the end of the gardening season nor the only time for indoor gardening. It is simply the beginning of another cycle. Growers have several reasons for gardening, whether it's a passion for plants, solace, entertainment, inspiration for art, personal satisfaction, religion, way of life, or something else. They rely on sheltering walls and sunny windows to make gardening possible—whatever the weather.

Newbie plant handlers might be skeptical about growing plants indoors. After all, plants usually thrive outdoors where there is plenty of sun, rain, and soil. But all outdoor gardens are at the mercy of seasons and weather, even in the mildest climates. Indoors, gardeners can control climate. Indoor gardens bring springlike blooms to the frozen North and re-create jungles in the desert. Gardening indoors is just the right antidote to cabin fever when winter keeps people housebound and flowering plants displayed around the house offers beauty and hope for spring.

Seasons and holidays

Plants provide long-lasting, welcoming gifts—cheerfully given alternatives from bouquets or bottles of wine. Beyond their beauty and fragrance, seasonal plants are symbolic connections to holidays and traditions. As one example, a holiday cactus cherished by generations of a family reblooms annually, serving as a reminder of those who have shared their penchant for gardening with heirloom cuttings. Poinsettias and tabletop evergreen trees add to winter celebrations with festive colors. Easter lilies trumpet spring; chrysanthemums brighten autumn gatherings. Flowering bulbs create season-to-season parades, from winter amaryllis to spring daffodils, summer lilies, and autumn crocuses. Plants enhance home interiors, from entryways to hearths, and from kitchen windowsills to dinner tables. They add color, fragrance, and timeless appeal to some of life's most memorable moments.

opposite, top right In late winter or early spring when the outdoor surroundings are typically dreary, a colorful tabletop garden sings a lovely song of spring color. Cheerful pansies, violas, and crocuses join hyacinths, daffodils, and pussy willow branches. **opposite, bottom right** Bromeliads create the festive ambience for a party—and a spark for table talk. Longer lasting than cut flowers, colorful plants are dramatic centerpieces. **opposite** Cans that once held imported tomatoes boast colorful labels and get extended use in housing a kitchen-window herb garden.

Enhance Health with Plants

Beautiful houseplants improve air quality in homes and offices.

Plants improve the quality of people's lives. At home and in public spaces, plants enhance mental, physical, and social health. As people spend less time outdoors breathing fresh air, indoor plants can be used to benefit health by boosting oxygen and moderating humidity to natural levels. Studies have shown that people who work with or near plants and flowers have lower blood pressure, demonstrated less stress, and have an increased sense of well-being. Research indicates that indoor spaces with plants can bolster serenity, reduce depression, soothe the senses, and evoke happy memories.

Natural air cleaners

In increasingly energy-efficient houses and office buildings, an unhealthy lack of fresh air and a preponderance of pollutants exists. The Environmental Protection Agency (EPA) ranks indoor air pollution as one of the top threats to indoor health. Synthetic furnishings and building materials that release toxic chemicals add to the problem. Hundreds of volatile organic compounds (VOCs) that form toxic vapors at room temperature are commonly used in plastics, cleaning products, and other household products.

Research by NASA has proved that an array of houseplants filters and significantly reduces indoor pollutants and particulate matter from the air. A NASA study suggests that as few as 15 houseplants can reduce the pollutants in an average home. One potted plant per 100 square feet can clean the air in an average-size house or office.

left Air-cleaning plants add more than welcoming greenery to a home. These include (clockwise top) peace lily, Chinese evergreen, and snake plant.

Toxic houseplants

Although most plants are harmless, some that are poisonous should be grown with caution, especially in households with children and pets. Avoid plants with spines or thorns, such as cactus, pygmy date palm, and sago palm. Other plants are harmful when chewed or swallowed, even though most must be consumed in large quantities to cause damage. Poisonous plants include Chinese evergreen, croton, flamingo flower, peace lily, pothos, schefflera, and splitleaf philodendron.

If you suspect that a child or pet has been poisoned by eating or touching a houseplant, call the 24-hour National Capital Poison Center at 800/222-1222, call your doctor or veterinarian, or go to an emergency room. Take a plant sample or photo with you to correctly identify the problem plant.

NATURE'S AIR FILTERS

These houseplants reduce common indoor air pollutants. Gerbera daisies and mums top the list of flowering plants most effective as air cleaners.

1. BOSTON FERN *Nephrolepis exaltata* is especially efficient at removing formaldehyde (found in carpet, fiberboard, and foam insulation) and adding humidity to indoor environments.

2. CHINESE EVERGREEN Research has shown that *Aglaonema* is particularly effective in filtering benzene (found in glue, spot remover, paint, varnish, and paint stripper).

3. DRACAENA *D. deremenis* 'Janet Craig' and 'Warneckii', *D. fragrans* (corn plant, shown), and Madagascar dragon tree (*D. marginata*) filter trichloroethylene from paint, varnish, and adhesive.

4. PEACE LILY *Spathiphyllum wallisii* is effective at absorbing benzene and trichloroethylene toxins commonly contributed by paints, lacquers, and adhesives.

5. PALM Effective at removing most indoor air pollutants. Choice plants include areca palm (*Dypsis lutescens*), bamboo palm (*Chamaedorea erumpens*), and lady palm (*Rhapis excelsa*).

6. SNAKE PLANT *Sansevieria trifasciata* 'Laurentii' is one of the best plant filters of benzene, formaldehyde, trichloroethylene, xylene, and toluene.

7. POTHOS *Epipremnum aureum* ranks high for removing formaldehyde, often found in insulation, paper products, household cleaners, and cigarette smoke.

8. ENGLISH IVY *Hedera helix* has proved effective at absorbing benzene (sources include cigarette smoke and car exhaust) and formaldehyde (commonly found in building materials).

9. SPIDER PLANT A single *Chlorophytum* in an enclosed chamber filled with formaldehyde removed 85 percent of the pollutant in a day in a NASA study.

10. WEEPING FIG *Ficus benjamina* works especially well at absorbing formaldehyde, toluene, and ammonia from indoor air.

CHAPTER TWO

Indoor Plant Decor

Heighten home appeal with potted plants,
arranged en masse or as a focal point.

Design Basics

Houseplants offer many different leaf types, textures, and colors. Mix, match, and blend to express your style.

Plants add personality to a home or office, dressing the spaces with decorative qualities. Diverse plant possibilities brighten dull corners with living forms, textures, and colors. For effective, long-lived appeal, vary shapes and sizes of leaves and flowers. Blending plants with interiors and existing furnishings with an awareness of design details ensures that plants fit into and enhance a room.

Form

Plant outlines or overall shapes create strong initial impressions. Combine a vertical (snake plant or pothos climbing a moss pole) with a mounded form (peperomia) to create drama and excitement. Plants vary in shape from arching, trailing, creeping, and rosette to bushy, dense, wiry, and frilly. Contrast or harmonize leaf shapes—from feathery, swordlike, and heart shape to grassy and serrated—to increase visual interest. When selecting a plant for indoors, consider form and how it will work in the room and with other plants.

Texture

Leaves and flowers create visual interest with endlessly fascinating textures. Highly textured begonias or aralias contrast with smooth-leaf peace lilies or cast-iron plants, creating a sense of depth and energetic dynamic. Planting ruffly primroses at the base of spring flower bulbs adds texture to the display as well as movement and intrigue.

left Grouping plants with varying forms as well as contrasting leaf shapes, textures, and colors creates a pleasing and natural effect.

Color

Of course you can match plant colors to interior decorating schemes. Also use color to create a welcoming and comforting home—playing with warm, exciting reds or cool, calming blues. Purple grabs attention; yellow and pink convey optimism. Colors clearly have emotional value. Soul-soothing green is the primary color of indoor plants, in hues of green as well as seemingly endless variegations (green marked with a pattern of white, gray, silver, yellow, red, pink, or purple). Choosing colors is part of the fun of gardening indoors. Color preferences will likely change with seasons, shifting from winter's white and icy blue to spring's bright pinks, purples, and yellows. Bold colors make beautiful, energized statements in summer's stronger light, yet gardeners often look forward to the transition to autumn's muted earthy tones and comparably quieter ambience.

Nature provides an array of leaf and flower colors and combinations. Pale or bright color foliage and flowers brighten dim areas; dark leaves and flowers stand out against a light background. When combining plants in a display, concentrate on the interplay between two or three colors, such as chartreuse, white, and pink. If you choose plants and containers that you like, focusing on your favorite colors, the result is often more a work of art than an act of nature.

BHG TEST GARDEN TIP — LEAF COLOR

Many foliage plants with colorful leaves need bright light to bring out the richness of color. When given insufficient light, some revert to all green.

PLAYING WITH COLOR

The power of color extends to pots and accessories. Go beyond leprechaun-green plants in terra-cotta pots. Paint pots to suit your style or use cachepots (decorative vessels) to hide utilitarian pots and set the tone for your decor.

BOOST COLOR
Foliage plants in low-light areas can display splashes of color. Gold bands on snake plant and red leaf margins on Chinese evergreen amp up the color.

CATCH LIGHT
Gather bottles old and new; blue and green are soothing. Add water and cuttings of coleus or other favorite plants.

ADD SHEEN
White and silver flowers and leaves blend well with other colors. Try cyclamen, star of Bethlehem, and kalanchoe.

REPEAT EFFECTS
Bold colors of matching-size containers create a pattern for a wall display. Leaf textures and shapes (Swedish ivy, baby's tears, snake plant) stand out.

Design with Plants

From tabletops to floor groupings, select plants from small to tall to create beautiful leafy vignettes in your home.

Indoor gardening presents continuous choices. A vast palette of versatile plants offers foliage and flowers to fit any design. Foliage plants may form the backbone of an indoor garden, while variegated leaves can add splashes of color. And flowering plants make an indoor garden spectacular. Combining foliage and flowering plants makes an impact.

All kinds of plants

Various plants have specific qualities that make them desirable for indoor gardens. Trailing and climbing plants can form verdant curtains, room dividers, or living walls. Tropicals grace indoor rooms with architectural forms as well as outdoor rooms during summer. Low-maintenance succulents and cacti have dramatic textures and forms. Like bonsai, they provide living art.

Adding plants

When acquiring a plant, look beyond aesthetics and consider where you will place the plant and how it might work with other plants in your home. One plant displayed by itself in an outstanding pot makes a stellar focal point. Repeating the same plant in multiple containers creates a rhythmic pattern that's sure to draw attention and enliven a room.

left Single plants—showy caladium and water lettuce—are combined in minutes as a serene desktop garden.

Grouping plants

Experiment to combine plants with varying sizes, shapes, and textures in an artistic display. Most indoor gardens evolve over time with periodic changes. Group plants as single plants in individual containers or as a mixed planting in an accommodating container. When combining plants in one container, choose only varieties that need the same cultural care. Select one plant—usually a larger one—to anchor the group. Let smaller and trailing plants reside at the perimeter of a grouping. Arrange a group of plants depending on how it will be viewed from one or more sides.

When designing a foliage-only display, focus on one shared quality of the plants—such as heart-shape leaves—to simplify the process and achieve a successful combination. Grouping very different plants in same-color pots or otherwise similar containers is effective, and the potted plants can easily be regrouped or displayed individually. Using similar containers for all your plantings unifies the decor.

As another option, massing multiples of one plant variety in one container also works well, especially for plants that appear skimpy on their own (such as geranium, Norfolk Island pine, or dracaena). Whether grouping a few small plants in a terrarium or multiple plants in a large container, allow space between plants for air circulation and growth. Grouped plants benefit from increased humidity among them.

BHG TEST GARDEN TIP — PREVIEW PLANTS

When shopping for plants at a nursery or garden center, see how combinations of plants work by placing them together in a shopping basket or cart.

GROUP ARRANGEMENTS

A grouping of plants does not automatically add up to a pleasing indoor garden. Choose and arrange plants according to simple design principles that suit your personal style and home decor, as well as the natural aesthetic qualities of the plant.

SHARED SHAPE
Heart-shape leaves unite this group, which includes a flamingo flower, arrowhead plant, begonia, and satin pothos.

COLORFUL ACCENTS
Grouping plants such as orchids gives them a strong presence. Small pots fit on gravel-filled trays. A mirror reflects the colorful display.

TABLETOP GARDEN
Make a long-lasting centerpiece in minutes using small potted succulents and tumbled stones—unconventional and attractive.

MIXING SINGLES
Herbs and tulips, potted singly in vintage containers for cohesion, form a pretty windowsill garden.

Find Your Style

Express your decorating style with indoor plants. Some deliver desert vibes. Others have the feel of the jungle.

Plants bring a room to life with color, texture, and vibrancy. Think of plants as living decorative accessories and use them to complement the design style of your setting and furnishings. Although the decorating scheme of any home is a personal expression, plant selections can work as finishing touches to make a stylistic statement. Minimalist, retro, eclectic—whatever your style, plants make a home more comfortable.

Architectural accents

Use plants in any room to enhance architecture. If you have a large open room with a vaulted ceiling, place a 5-foot-tall palm on a pedestal to give the illusion of a large plant without overwhelming the space. Other architectural features that call for plants are alcoves, ledges, fireplace mantels, built-in bookcases, and windows. Wherever you display one plant or many, attention will be drawn.

Container choices

Express style with the selection of pots. Use building materials, wall colors, and fabrics as guides, or go with neutral containers that will work in any setting. Paint terra-cotta pots any color, or use only metallic cachepots.

top left Plants add lively charm to any home. Succulents in diverse containers create artful accents that are rich in texture. ***bottom left*** Display plants with collectibles, such as vintage birdhouses. ***opposite, far right*** Flowering plants, such as cyclamen, mix with any style.

Minimalist

This contemporary style simplifies with zenlike focus: less clutter and more essential beauty, elegance, and calm. Low maintenance is a bonus.

PLANT POSSIBILITIES

Aloe

Sedum

Snake plant

Casual

Call it cottage, country, or garden style, a plant collection helps combine comfortable-yet-chic decorating with other natural elements.

PLANT POSSIBILITIES

Bromeliad

Moss

Orchid

Modern

A spare architectural setting with sleek lines calls for single plants as living sculpture and striking effects.

PLANT POSSIBILITIES

Agave

Kentia palm

Schefflera

Traditional

Composed with attention to bold and graceful lines, this style balances formal embellishment with livability.

PLANT POSSIBILITIES

Fern

Peace lily

Topiary

Regional

Desert surroundings offer a cue for decorating with native plants, a fitting color palette, and local stone and wood.

PLANT POSSIBILITIES

Citrus

Ornamental chile pepper

Prickly pear cactus

Urban

Loft or condo living calls for the personal flair of bold colors, strong patterns, and lush greenery.

PLANT POSSIBILITIES

Croton

Dracaena

Zeezee plant

Make Room For Plants

Add life to every room with flowering and foliage houseplants. Select species based on each room's growing conditions.

Improving the appearance of a room by incorporating plants can be thought of as interior landscaping or plantscaping. Each room in a home has a different function, and its practical use combined with the growing conditions determines which plants could do well there. Growing conditions usually vary even in different parts of a room, depending on location of windows, doors, and air vents.

Successful placement

Not everywhere in a room is good for plant growth. For example, areas near doorways are subject to cold air that can devastate some plants. In entryways and hallways, choose hardy plants that fit the space. Plants should not obstruct stairs or doorways. Wherever you place plants, ensure that they can be easily watered, and place an impermeable barrier under any container to protect surfaces from moisture and scratches. In living rooms where families gather and entertain, plants should enhance rather than clutter. Lack of humidity and low light are main obstacles to growing plants in a living room. Often overlooked rooms—the master bedroom and guest rooms—can be ideal places to raise plants.

left If an entryway has a window, there will be enough light to keep herbs going for a few months. Trading them for seasonal bloomers keeps the display fresh. *opposite, below right* A trio of African violets in similar size and shape pots creates a beautiful windowsill vignette.

Entryway
Colorful, pretty plants create a welcoming first impression. A window enhances the light.

PLANT POSSIBILITIES
Chinese evergreen
Ivy
Ti plant

Living Room
Light level is typically medium, and there's room to group plants with ornamental objects and furniture.

PLANT POSSIBILITIES
African violet
Cyclamen
Pothos

Kitchen
Typically a good place for plants, kitchens offer easy access to water and higher humidity than other rooms. A window is ideal.

PLANT POSSIBILITIES
Culinary herbs
Lettuce
Succulents

Dining Room
Living centerpieces and floor plants are ideal. Choose short tabletop plants to allow uninterrupted eye contact. Avoid heavily scented plants.

PLANT POSSIBILITIES
Begonia
Oxalis
Peperomia

Bathroom
Take advantage of high humidity near sink, tub, or shower and add a plant where space allows.

PLANT POSSIBILITIES
Bromeliad
Fern
Orchid

Workspace
In a home office or spare room with multiple windows and ample light, most plants will be content.

PLANT POSSIBILITIES
Orchid
Scented geranium
Swedish ivy

left Displaying herbs, ivy, wheatgrass, and forced bulbs at different levels makes a room more inviting and interesting. Potted plants work best when set away from foot traffic. **above** Lily-of-the-valley and other fragrant plants merit a place of honor where their pleasant scent will waft across a room.

Siting Plants

Use plants to enhance a room, separate a space, or camouflage a less-than-desirable view.

Once you have determined how much space for plants your home allows, it's time to select specific plant-friendly locations. If you have an area you want to accentuate or a bare space to fill, start there to decorate with plants. On a windowsill or coffee table, shelf or pedestal, use plants to create privacy, decorate for a party, mark an entry, keep produce handy, or add flowery fragrance. Strategically place plants to camouflage less-than-desirable views or separate a large room into inviting areas for multiple uses. In a room with unused space that's too small to accommodate a functional piece of furniture, or too large to be ignored, put plants to work there.

Location, location
Windowsills are among the most natural perches for indoor plants because of the optimal light conditions. Choose small potted plants that will stand steadily on a sill, or mount a deep shelf on a window frame to hold big pots. Vining and trailing plants work well in hanging planters, suspended to screen or frame the view outdoors.

Any plant that is low growing and interesting when viewed from above makes a good tabletop plant. Group a few small plants for an intriguing tabletop display, or focus on a single spectacular plant. When a plant outgrows a tabletop, move it to a bookcase or shelf where it might cascade or best show off its maturing form.

Architectural plants
Most furniture is not designed to fit into corners, but a plant or a group of plants brings an empty corner to life. Large plants with interesting forms (weeping fig, palm) or large leaves (alocasia, philodendron) command a place of prominence. Place an architectural plant where its silhouette will stand out against a wall.

PLACES FOR PLANTS

Feature a prized plant—or a grouping of plants—prominently where it will create a stir and everyone can enjoy its presence.

FLOOR PLANT Place a plant, such as an aloe, which is too large for a tabletop or grouping on the floor to produce drama from its size alone.

PLANT STAND Raising plants on a tiered stand gives the grouping greater impact. The plants ('Song of India' and 'Janet Craig' dracaenas and Red Flame Vriesia) create an effective display.

WALL ART Staging a Phalaenopsis orchid on a wall sconce gives it the spotlight it deserves. The ribbon and ornament add a holiday note.

Special Effects

Add drama to interior settings with houseplants that infuse the space with color, texture, and distinctive shapes.

Situate plants in your home according to available space. Scale and proportion of rooms are important considerations in what plants you select. If you want the drama of an extra-large potted tree, the ceiling height will make it possible—or not. Make a bold statement with a 9-foot-tall weeping fig tree in your living room only if the plant and its container fit together in the room proportionately and do not overwhelm the space or each other.

Big and little plants

Both large and small plants create attractive focal points. Take advantage of plant size in relation to the scale of the setting. Big or tall plants have the most distinctive effect when placed on the floor away from other plants and furniture. Regard them as furniture or sculpture that adds style, beauty, and even function in your home rather than just taking up space. Small plants draw attention to their diminutive stature. Group a few small potted plants on a tray or in a dish, terrarium, or other eye-catching container, then situate them where they're certain to get a closer look. Lining up a few ivies in silver pots on a mantel or gathering several orchids in a flat basket on a side table will have more impact than scattering single potted plants around a room.

left Fiddleleaf fig (*Ficus lyrata*) is one of many trees that will live indoors for years and offer bold presence. Other tree-size plants include Norfolk Island pine and African milk tree (*Euphorbia trigona*). ***opposite, top left*** Pots of blooming kalanchoe form a contrasting skirt at the base of a tall cactuslike African milk tree. Holiday cactus or small poinsettias would also work well.

When space is limited, locate plants where they will add color and character without competing for room. Several well-placed 2- to 3-foot-tall plants can freshen room decor. A selection of seasonal beauties—kalanchoe, begonia, cyclamen—brightens clusters of plants, whether added individually or massed.

Balance

Ultimately, selection and placement of plants depends on what appeals to you and makes a positive impact. Any arrangement that includes plants and other objects should be balanced either formally or informally. Using two single potted plants to flank a feature, such as a fireplace or doorway, has a formal effect. Topiaries, tree-form plants, and upright cacti work well in formal schemes. Informal designs are asymmetrical and freely contrast the forms, colors, and other design qualities of an odd number of plants.

SPECIAL ARRANGEMENTS

Displaying plants attractively may take several attempts to find just the right plant or combination for the location.

FOLIAGE FOUNTAIN
Although this stone fountain no longer pumps water, it provides a creative framework for air plants (*Tillandsia* spp.).

SEASONAL SENSATION
A trio of braided-stem azaleas forms an outstanding formal centerpiece on a dining room table.

LIVING TREASURE
Tiny air plants (*Tillandsia caput-medusae*) deserve to be showcased where they will receive notice—as in this small box.

Ideal Environment

Map out areas of your home to understand where plants will thrive and excel.

The best place for any plant is where it will receive the light it needs to survive and thrive. Placing plants where they contribute to the decor is only part of a gardener's task in making a successful indoor garden. If you have any doubt about the suitability of a spot for a particular plant, experiment. Just be aware that the plant might not grow well or remain alive for long. The more dependable approach starts with knowledge about each plant's environmental needs for light, humidity, temperature, and air circulation.

Habitats for plants

Knowing the plant's place of origin will help map out areas of your home that offer conditions conducive to its well-being. In natural habitats, plants have evolved and adapted to specific climatic and soil conditions. As a gardener, you can match plants to comparable conditions in your home and adjust the level of heat, humidity, or air movement as needed.

Microclimates

Just as plant needs for light and water change with seasons, the conditions in homes vary from month to month, room to room, and within some rooms. Homes have microclimates where windows, vents, and other features affect the environment, making it warmer or cooler or more or less humid. Use these microclimates to benefit plants. For instance, north windows keep a room cool during winter and are the perfect place for

left Sheer curtains diffuse the bright light directly in front of a south-facing window, making it ideal for African violets, begonias, and maidenhair ferns. **opposite, top left** Warmth and bright light are crucial for the success of herbs including rosemary, sage, thyme, parsley, and chives.

LIGHT CONDITIONS

The amount of light available to a plant indoors is critical to its health. The classifications of low, medium, and bright natural light provide guidelines for placing plants.

plants that rest between late fall and early spring. Plants native to arid climates do well near heat vents; plants that hail from the tropics appreciate high humidity of bathrooms or kitchens. Placing a plant on a mantel or other spot near a fireplace may appear pleasing to the eye, but the plant should be moved away from the area if a fire is started to avoid the risk of being scorched.

BRIGHT LIGHT Plants receive 4 or more hours of high or bright light in an unobstructed window that faces south or southwest.

Plant Possibilities
Cactus, Citrus, Croton, Flowering maple, Hibiscus, Jade plant (shown), Orchids, Peperomia, Purple passion plant, Succulents

MEDIUM LIGHT In an east or west window, this site receives a few hours daily of early-morning or late-afternoon sun.

Plant Possibilities
African violet, Begonia, Dracaena, English ivy, Ferns, Fig, Nerve plant (shown), Peperomia, Schefflera, Spider plant

LOW LIGHT In a place well away from a north-, east-, or northeast-facing window, plants get some light but not direct sun.

Plant Possibilities
Cast-iron plant, Chinese evergreen, Dieffenbachia, Dracaena, Lady palm, Parlor palm, Peace lily, Philodendron, Satin pothos (shown), Snake plant

CHAPTER THREE

Plant Care

The fundamentals of growing healthy plants indoors include tried-and-true techniques as well as updates to further your know-how.

The Right Light

An essential for all living things, light is critical for the health of plants.

Light is vital to plants. The ability for plants to grow, maintain health, and produce flowers depends on light. Specifically, plants need light for photosynthesis—which produces the food and energy necessary to keep them alive—as well as for hormone production, to induce flowering. Ideal light levels vary by plant species and sometimes cultivar.

If a plant is sited in more or less light than it needs for optimum growth, it will be stressed and prone to problems, such as weak growth, minimal flowering, disease, and pests. Plants usually show signs of light imbalance. Symptoms of insufficient light include stretching or leaning toward a light source, growing sparsely or becoming spindly, losing foliage color or variegation, and producing smaller leaves. Symptoms of too much light include whitish scorch marks on foliage, wilted and shriveled leaves, or bleached leaves.

Light levels

Before placing a plant indoors, identify the amount of light it will need and the location where that light is available. You may need to experiment until you discover where the light in your home best suits a plant.

Plant tags usually state what level of light a plant needs: low, medium, or bright. Some plants thrive in a range of light levels. Low light may be enough to keep a plant alive, but not enough to promote flowering. Several hours of bright light in a west window sustains herbs and succulents, but they'll grow better in the full strength of midday sun in a south window. Bright, indirect light can often be found within 3 feet of a sunny window. Use a light meter to gauge the amount of available light, or look for shadows: The brighter the light, the stronger the shadows.

Some plants are more forgiving of light levels, including the intensity, quality, and duration of sunlight, which changes with the seasons. Winter light is less intense than summer light, but it extends farther into a room because the sun is lower in the sky.

above Most homes, even the smallest apartments, offer a variety of light conditions. Each setting is an opportunity for different plants. Foliage plants usually need less light than flowering ones.

Intensity: The closer a plant is placed to a window, the more intense the light. Bright, direct light in a south- or west-facing window suits succulents and seedlings, but it could toast African violets and bromeliads. Sheer curtains diffuse bright light and help prevent sunburn (yellow to brown scorched leaves). Indirect light is usually better for indoor plants that require low to medium light. Moving plants away from a window reduces the light they receive. A plant that needs moderate or medium light will thrive in an east-facing window but should be set a few feet away from the intense light of a west window.

The intensity of direct light coming through a window also varies with seasonal changes. In summer, when the sun is high in the sky, an overhanging roof can block its rays. The winter sun is lower in the sky, but its light is less intense. Outdoor elements, such as trees and neighboring buildings, can reduce light coming indoors; reflective surfaces, such as light-color walls and mirrors, increase it. To ensure that plants get the optimum light to maintain health, relocate them periodically.

Quality: Sunlight is the best source of full-spectrum light for plants. Supplement natural light with artificial light to provide plants with the amount and quality of light (wavelengths in the red-blue spectrum) they need to thrive.

Duration: Most indoor plants need from 8 to 16 hours of light daily. Too many hours of light slow plant growth. Too few hours cause elongated shoots and thin, easily damaged leaves. Many plant blooms are triggered by length of day and night cycles.

Supplemental light: When natural light is inadequate, provide supplementary light to help plants fare well, especially during winter. Seedlings and flowering houseplants, such as African violets and begonias, benefit from an added light source. Most plants respond best to normal cycles of 12 to 16 hours of light and 8 to 12 hours of darkness when given supplemental light. Place the light within 4 to 6 inches of plant tops. Fluorescent lights are available in various shapes and sizes, and as cool- or warm-white tubes, at hardware stores. The lights are economical and remain cool to the touch, allowing plants to grow close to the tubes without damage. Costly fluorescent grow-lights mimic full-spectrum light. Other more intense and expensive options for avid gardeners include metal halide lights and high-pressure sodium bulbs.

SUPPLEMENTAL LIGHTING

To ensure plant survival, it is sometimes necessary to provide artificial light. Indoor gardeners have several options to give plants the light needed.

SHOP LIGHT A fluorescent light fixture holds one cool-white tube and one warm-white tube that produce sufficient light for seedlings indoors. An electrical timer turns the light on and off, providing a consistent and convenient light source for a sufficient time.

LITTLE LIGHT Where natural light is lacking, a supplemental LED (light-emitting diode) light stick casts a refined spectral wavelength to meet individual plant needs. This efficient light source uses little power, produces little heat, and lasts longer than most bulbs.

EASY FIXTURE A clamp-on fixture outfitted with a grow bulb provides an attractive means of shining full-spectrum light onto plants in an otherwise dark corner. Adding a grow bulb to a small table lamp works too.

Watering

All plants need water. Plant type, season, and container determine each individual plant's watering needs.

Water can be a cure-all or a killer for plants, depending on when and how it is applied. No single rule applies for watering, because there are many variables. A plant's needs change with the seasons. Plants absorb more water when the light is bright, temperature is high, or humidity is low.

The thirstiest plants include those with flowers, large leaves, tropical or marshy origins, or large root balls in a confined space. Plants in unglazed terra-cotta pots, baskets, and wooden vessels need more frequent watering than those in plastic, metal, or glazed clay pots. A lightweight soilless potting mix dries out faster than heavy potting soil. Fast-growing plants (ferns, dracaena, palms) need more water than slower-growing ones (succulents, snake plant). Small pots dry out faster than large ones.

When to water

Check the soil of potted plants twice a week to learn about water needs. Feel the soil before watering or use a moisture meter. Using your finger as a dipstick, poke it into the soil 1 inch (to the first knuckle) to determine whether the soil has begun to dry; 2 inches (to the second knuckle) to feel for moisture. A plant's need for water is individual. You will learn to feel the difference in soil that is dry, damp, moist, or wet, and when water should be added to maintain a particular level of moisture or dryness.

Underwatering and overwatering can result in the same symptoms: wilting, pale, lackluster foliage. Without enough water, a plant becomes stunted; leaves turn yellow or brown. If you water before a plant wilts and suffers permanent damage, it may recover. With too much water, roots can't function normally due to lack of oxygen in the root zone, and the plant suffocates.

left Watering deeply and less often is good overall strategy for keeping houseplants well hydrated.

Over time, you will discover which plants get by with weekly watering and which plants need more- or less-frequent hydrating. Plants usually need water more often in spring and summer, when they're actively growing, than in winter. Water in the morning, if possible, to give wet foliage a chance to dry during the day and minimize disease. Early watering provides a chance to see whether excess water has pooled in a saucer and needs to be poured off.

How to water

When watering a potted plant, pour water over the entire soil surface and let it soak into the soil. Add more water until it starts to seep out the bottom of the container. This practice thoroughly moistens soil and leaches excess salts from the soil. Allow the soil to soak up any excess water, then pour off the remainder to avoid letting a plant stand in water, which can damage the roots. Self-watering pots (with a water reservoir) and capillary mats work the same way as bottom watering, wicking moisture into the soil. They work well for plants that need to stay evenly moist and during times when you will be away from the plants.

Types of water

Tap water suffices for plants, unless they are sensitive to hard (alkaline) or soft water (which contains salts). Rainwater agrees with plants, unless air pollution in the region causes acid rain. Use slightly warm or tepid water for plants to avoid shock and possible root damage from extreme temperatures.

BHG TEST GARDEN TIP — WATER METER

For plants that are finicky about watering, use a moisture meter to measure the water content of the soil. Meters are inexpensive and take the guesswork out of when to water.

HOW TO WATER

You'll find more than one way to water plants indoors. The best methods saturate soil adequately without spilling or leaking water on furniture, floors, and other surfaces.

WATERING CAN Use a watering can with a long, slender spout to easily guide water to all sides of a pot and avoid spills. Watering cans come in a range of sizes. A small can serves one or two plants; a larger can saves trips to the sink when watering multiple plants.

WATERING BULBS These decorative devices provide a slow, steady supply of water to a plant, directing moisture to the roots, where it is most needed, and keeping soil evenly moist. Various forms are available. The bulbs shown are blown glass.

BOTTOMS UP To prevent water marks on leaves, water from the bottom. Fill a saucer or reservoir with water and allow the soil to wick it up through the bottom of the pot.

REHYDRATE SOIL If dry soil has pulled away from the pot, immerse the entire pot in a bucket or sink of tepid water. Let the pot soak for 20 minutes, then drain.

Success Factors

Humidity, warm temperatures, and good air circulation help plants feel more at home inside your home.

Indoor conditions affect plant well-being and survival. Conditions can be tough on plants, especially when soil dries out quickly during winter (low humidity) and summer (excessive heat). When these conditions prevail, help houseplants succeed with varieties that can survive periods of drought. Also keep in mind that small pots dry out faster.

Humidity

Moisture in the air is just as important to a plant as moisture in the soil. Warm air holds more water, so humidity tends to be higher in summer, although air-conditioning dramatically lowers humidity. In winter, indoor humidity tends to be lowest when heating systems dry the air. Generally, the colder the air outside, the drier the air indoors. The humidity in most homes varies from 20 percent in winter to 65 percent in summer, and from room to room. A hygrometer is a device that measures humidity. Devices that measure humidity and temperature are widely available.

More often than not, plants need increased humidity. Low-humidity problems are intensified if the soil has dried or if the plant is in a draft or flooded with unfiltered light. Methods for boosting humidity include running a humidifier, setting individual pots or groups of plants on a pebble tray, and planting in a terrarium. Group plants to allow leaves to release moisture (transpire) and raise humidity, which benefits more than a single plant. Double-pot plants, setting a terra-cotta potted plant inside a larger pot and filling the space between the pots with damp sphagnum moss; keep the moss damp by watering it. Although misting plants rinses the leaves, it promotes disease of some plants and does not effectively raise humidity. The higher humidity of

left A pebble tray is an easy way to boost humidity. Add pebbles to the saucer of a single potted plant or fill a large tray. Cover the gravel with water, but avoid letting pots and plant roots stand in water, which promotes rot.

rooms with running water (kitchen, bathroom, or laundry room) suits peace lily, African violet, ferns, and other plants from the humid tropics, as long as they also get enough light.

Temperature

Year-round temperatures between 60 and 75°F (and cooler at night) in typical homes suits most plants. Interior temperatures vary from season to season, room to room, and within a room. Take advantage of these differences by matching plants to the conditions and moving them as needed. You'll know within a week or so whether a change has improved a plant's health. Most plants are sensitive to a sudden drop in temperature. An extreme change can cause leaf damage or drop, or even plant death. Buds and flowers especially need warmth.

Microclimates exist within a room. A sunny, unheated porch or place close to a window in winter provides a cool microclimate that suits cyclamen, azaleas, and some orchids. A chilling draft near a window, door, or entryway can injure some plants enough to cause leaves to droop, turn black, or drop off; but a spider plant or schefflera will appreciate a cool location. Cacti and succulents are among the few plants that are happy in a home's warmest places: near a heat vent and on sunny windowsills in summer. Warm air rises, making a toasty place on top of a bookcase; cool air sinks, so the air near the floor is cooler.

Air circulation

Air movement around plants helps evaporate moisture from leaves which prevents disease, promotes growth, and keeps some insect populations in check. Open a window during mild weather to increase air movement. Keep air moving in winter with a ceiling fan or a small fan placed near plants.

PLANT SIGNALS

Plants demonstrate when the conditions are not suitable. They show symptoms of stress that signal a change is needed in their environment or care.

HUMIDITY

Too little Leaf tips brown and shrivel as on this peacock plant; leaf edges turn yellow. Buds fall off and blossoms wither.

Too much Plants are more susceptible to rot, mold, and mildew.

TEMPERATURE

Too cold Leaves curl, turn black or brown (shown), wilt, and fall off; buds drop when conditions are too chilly. Plant tissues, including roots, can succumb to cold.

Too warm Lower leaves wilt and turn brown. Flowers die quickly. The plant produces small leaves or weak, leggy growth in good light.

AIR CIRCULATION

Too little Fungal disease (powdery mildew, shown on rosemary) or insect problems (whiteflies) may proliferate.

Too much Wind shreds leaves, dries soil, and shakes off buds or flowers.

Selecting Healthy Plants

Starting out with leafy, lush, and robust plants is the best way to ensure long-term success in indoor gardening.

Acquiring indoor plants entails as much thoughtful selection and careful shopping as other purchases that will last for years. When you buy plants, find reputable local greenhouses or garden centers, or order by mail from a specialty catalog or website. You might find an irresistible plant at a grocery store, botanical center, plant society sale, or another opportunity. Shipping plants is risky. Dependable sources will ship plants when conditions are right and will guarantee to replace damaged plants.

Selecting a healthy plant

Examine plants before buying and pick the healthiest, lushest, most colorful, dense, and full plants you can find. New growth is a good sign; recent heavy pruning is not. Be wary if some plants in a display look good and others do not, especially if some are wilted. Closely inspect plants for indications of pests (webs, sticky residue, cottony masses) or disease (mushy, discolored, or distorted parts). A yellow leaf or two is not cause for concern, but lots of fallen leaves or extremely dry soil indicate stress and could signal inadequate care. Look at the undersides of leaves and poke into the soil for clues. Avoid plants that have weak or spindly growth, crawling or flying insects, or sour- or rotten-smelling soil. A discounted plant is not a bargain if it is sick.

left Choose new plants carefully. Be aware of a plant's potential size and needs, or you may find that cute little plant you couldn't resist quickly grows into a plant too large for the space or its needs prove difficult to fulfill.

Be certain of your purchase. If the plant does not have a tag that lists its name and growing requirements, ask someone on staff to provide information. Ensure that the plant suits the conditions you have to offer it at home, especially in terms of light and space. Young, small plants cost less than large ones; they also travel and acclimate with less stress.

Transitioning plants

When you acquire a new plant, remember that it will be vulnerable to changes in its surroundings, whether you transport the plant or it is shipped. Do what you can to give the plant extra-tender care in the process. Make sure it is well-packaged for travel and wrapped to protect it, especially from cold temperatures, between the point of purchase and your home. When transporting plants, avoid letting them sit in a hot car for an extended period: In winter, heat the car and place plants near the heater, not in the trunk. Once new plants—including mail-order purchases—arrive at your home, unwrap them immediately. Set a new plant in the sink and water it thoroughly.

Isolate a new plant for a few weeks in a room away from other plants, if possible, allowing the plant to recuperate from the stress of environmental change. At the same time, ease the plant through its adjustment to your home. Place it in high light for a week or so, then gradually move it to locations with less light if its intended place has medium or low light. Avoid overwatering during this stage.

BHG TEST GARDEN TIP
CHANGE AVERSE
Some plants (croton, weeping fig) are so sensitive to change that it's best to place new residents in their permanent spots and acclimate them there. Tougher plants (dracaena, philodendron) can be acclimated gradually and moved to different locations around your house in the process.

NEW PLANT SUCCESS

Follow these tips to help with the best plant selection and to ensure that plants adjust well to their new digs.

LASTING BLOOMS When choosing a flowering plant, buy the one with lots of buds—it will reward you with longer-lasting blooms. A plant in full bloom will soon fade and may or may not bloom again.

KEEPING TRACK It's easy to forget pertinent information about a plant: its name, source, and date of purchase. Use a journal or file to store details, and save the receipt to help track your plant's success or obtain a refund.

WELCOME HOME Help acclimate a new plant by nurturing it in highly humid conditions for 2 weeks. Set the plant in indirect, bright light and tent it with a plastic bag. Poke a hole or two in the bag every few days.

Choosing Containers

The right container can enhance the beauty of any plant. But pot selection involves more than just aesthetics.

As homes for plants, containers provide adequate space for roots to develop and plants to grow and flourish. They come in an array of colors and styles, which makes choosing a container almost as much fun as selecting a plant. It is a matter of personal preference whether to coordinate plants and pots with your home style for decorative results. If you acquire quality containers that can withstand rough handling and outdoor conditions, then they—and the plants they hold—can be moved outside during summer.

Choosing containers

Consider container size, weight, and shape in relation to the plant. To appear balanced, the container should be no taller than one third of the plant/pot combination. Selecting a pot 2 inches wider than the plant's root ball gives the plant the right amount of growing room. An oversize pot not only appears disproportionate, but it also holds more moisture than the plant needs. Many plants need cozy quarters to grow well. Small containers can become a source of frustration when plants outgrow them, and the plants will require watering more often. Heavy containers stabilize contents, preventing toppling of top-heavy plants. The weight of large pots filled with soil, plants, and water limits portability. Light-weight pots are easily portable. Balance an upright plant in a tall pot, a mounding plant in a square pot, and a trailing plant in a V-shape pot. Experiment with combinations of plants and shapely pots that appeal to you.

Drainage

Good drainage is crucial. Survival of potted plants depends on drainage holes or other means of releasing excess water. Without adequate drainage, plant roots can suffocate and rot. If a container does not have at least one drainage hole, drill one.

above Containers accent colors and styles of homes, whether new or old and even whether they hold plants or not.

Cachepots

In lieu of drilling into a container, use it as a decorative cachepot, hiding the nursery pot inside. When watering a plant in a cachepot, pour off any excess water that is not absorbed in the potting mix and that drains from the nursery pot. When using an ornamental cachepot made of wood, wicker, metal, or another material that could be damaged by moisture, set a plastic saucer under the nursery pot to protect the cachepot from damage. When hiding an ordinary nursery pot inside a cachepot, conceal the rim of the plastic pot under a thin layer of preserved green moss, long-fiber sphagnum moss, or Spanish moss.

Material matters

Key to a plant's success is the ability of the container to hold soil and moisture, which depends on its construction material. Porous materials, such as terra-cotta and wood, dry out faster and are good for plants that prefer soil on the dry side. Nonporous materials, including plastic and glazed ceramic, hold moisture more evenly and longer. Well-made containers stand up to watering and weather. Unless treated with a waterproof sealant, wet wood rots and metal corrodes.

TYPES OF CONTAINERS

Containers come in many styles and materials.

1. RECYCLED A new generation of eco-friendly containers is made from recycled plastic, paper, rubber, and other materials, saving resources from landfills.

2. TERRA-COTTA Affordable terra-cotta is weighty and porous, washable and breakable. High-fired terra-cotta is most durable; hand-thrown and Mexican pots are more fragile.

3. GLAZED CERAMIC Choose from an array of colors and patterns that can tie into a room's decor. Handle pots carefully to avoid chipping. Ceramic pots are weighty.

4. BASKET Sturdy and attractive options include an array of woven materials. Natural materials require more protection from moisture damage. Use liners to protect the containers.

5. VINTAGE Planters from another era have charm and collectability. Pots with chips, peeling finishes, or other imperfections may be available at a discount.

6. METAL Galvanized metal, copper, and zinc offer versatile style and long life. Shiny metal surfaces fade into subtle patinas unless they're powder-coated, painted, or polished.

7. WOOD Line wooden containers with plastic before planting or use them to hold pots. Choose rot-resistant cedar or wood that has been painted, stained, or sealed.

8. PLASTIC Plastic is lightweight, relatively unbreakable, and easy to clean and store. Prices vary widely with finish and quality; some plastics are long lasting.

9. SELF-WATERING A built-in reservoir allows plants to draw moisture as needed.

10. REPURPOSED From coffee cans with rusty patina to desk drawers, beach buckets, or wastebaskets, any item can become a container. Drill a drainage hole or use as a cachepot.

Plant stands give legs to a container, protecting the floor from water damage and using vertical space to display plants.

42

Displaying Pots

Plants are assets to indoor decor, and they require special care for how and where you show them off.

Little else will dampen enjoyment of an indoor garden more than a leaky container and moisture damage to surfaces. Few surfaces are waterproof, and it doesn't take long for water to stain or ruin carpet, wood flooring, furniture, windowsills, or countertops. Take a preventative approach to moisture damage. Avoid placing plants on precious furniture, such as a piano or an antique.

Saucers and trays

Always use a saucer or tray under a pot to catch excess water that drains from the container. The saucer or tray should be wider than the pot's widest diameter—so water can drain into the saucer or tray without overflowing.

Also protect the surface under the saucer or tray from moisture that sweats, seeps, or spills and collects enough to cause damage. Use a large ceramic tile, cork coaster, or trivet that will raise the saucer or tray off the surface and allow air to keep the surface dry. Even a cachepot, ceramic pot, or other container that you think will hold water should have a saucer. Clear plastic saucers are widely available and inexpensive for use inside a cachepot or basket, or as a liner for a terra-cotta or glazed saucer.

Other options for raising pots above surfaces and preventing damage include pot feet or risers. Instead of a saucer or tray, repurpose vintage plates, cross sections or slabs of wood, or remnants of a stone countertop. Or have glass cut to fit a tabletop, windowsill, or other surface to protect it from potential water damage and scratches. Have the glass edges ground smooth for safe handling.

Use felt to make coasters for potted plants with saucers. Cut a piece of felt the same size as the bottom of the saucer and tack it in place using white glue. Coasters will prevent saucers and potted plants from scratching the surface of furniture.

WAYS TO HOLD POTS

Give indoor gardens a lift, using pot holders of any kind to raise containers off a floor, table, or other surface.

1. STEPPED-UP DISPLAY A small painted step stool is an effective stage for small pots on a table. This flowering garden includes calla lily, hydrangea, lily, and kalanchoe.

2. PLANT CADDY Setting a large, heavy potted plant on a caddy with casters makes it easy to move around the house or outdoors. Some caddies feature a built-in saucer.

3. STANDING PLANTER A collection of ferns stands out in this wicker planter. The grouping of two maidenhair ferns and a rabbit's foot fern in 6-inch pots fills the painted wicker with lush greenery.

4. SAUCERS GALORE A wide variety of saucers, drip trays, and coasters is available in many styles and materials, from terra-cotta to plastic, ceramic, copper, stone, and cork.

Plant stands

Some plant stands have feet, others have legs, and all raise potted plants above the floor. Consider using a plant stand with multiple shelves and a tiered design to display a container garden at varying heights and put plants within easy reach. Made of metal or wood, the ideal plant stand moves outdoors with your plants for summer and weathers the elements. Beware of setting large and heavy containers on the top shelf of a tiered stand, making it top-heavy. A tall or tiered plant stand or baker's rack may need to be secured to a wall.

Potting and Repotting

If you own a plant long enough, it will eventually need to be repotted. Fortunately, it's an easy process.

Any young plant growing in a small pot will need to be repotted before long. Many mature plants seldom need repotting. Some plants bloom and perform better when roots are snugly enclosed or even crowded in a pot. When the roots have filled a pot so tightly that they begin circling the root ball or the bottom of the pot, the plant has become pot-bound or root-bound and needs more growing room in a larger pot.

Time to repot

Plants provide clues to help determine the proper time to repot them. Lift the pot and see whether roots have begun growing out of the drainage holes. The roots may be so densely packed that little soil is left and water runs through the pot. To check, tip the pot on its side, gently dislodge the plant, and slide it out of the pot to examine the root ball.

A few plants grow less vigorously and won't bloom when they become pot-bound. After blooming for two years in the same pot, holiday cactus, wax plant, clivia, and spider plant benefit from repotting just after a bloom period. Ferns, palms, amaryllis, and some other plants that don't like to be disturbed by repotting or do best in close quarters benefit from fresh soil each year. Carefully scrape off the top inch of soil and replace it with new soil.

Other signs that a plant needs repotting: the plant has become too top-heavy for its container and topples easily, it wilts soon after watering, it develops new leaves with decreasing size, it displays yellowing lower leaves, or it has become infested with insects.

Microclimates

Attention to detail when potting or repotting ensures a strong, healthy plant. A day or two prior to repotting, thoroughly water the plant to prepare it for transplanting. To repot a plant, tip the container on its side and carefully slide out the plant. If it doesn't come out easily, run a table knife around the inside of

HOW TO REPOT A PLANT

When possible, repot a plant outdoors on a work surface covered with newspaper. Or repot on a waterproof countertop. When adding new soil to the plant, fill lightly; do not pack it into place—roots need air space for water and growth.

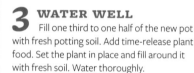

1 LARGER POT
Choose a new pot 1 to 2 inches larger in diameter than the root ball of the plant, giving the plant room for root growth to the sides and bottom of the pot, as well as space for watering above the soil line.

2 LOOSEN ROOTS
If the plant won't budge from its old pot, gently squeeze the root ball to loosen the roots. If the roots are tightly bound, push a finger into the mass and tease the roots loose.

3 WATER WELL
Fill one third to one half of the new pot with fresh potting soil. Add time-release plant food. Set the plant in place and fill around it with fresh soil. Water thoroughly.

the pot. If the soil is depleted or infested with insects, take the plant outside and rinse the root ball with a garden hose. Examine the roots for any problems, prune out damaged or diseased areas, and repot the plant.

When potting, set a plant at the same soil level as its previous planting. Burying any portion of a plant too deeply can cause rot. If grouping plants in a pot, first arrange them in the nursery pots. Once satisfied with the grouping, remove the plants from nursery pots, setting the largest plant in place first and the smallest plant last. Fill in between plants with soil, leaving 1 to 2 inches of watering space between the soil surface and the rim of the pot.

left This Chinese evergreen completely fills its container with a mass of roots and stems. The plant has become so large for the pot, it tips over easily. Time for repotting! **above, top** Repotting your plants gives you a chance to change containers. Just make sure that the new container has adequate drain holes. **above, bottom** If the roots are tightly massed and circling the root ball, the plant is root-bound and needs to be repotted. Gently loosen the roots before repotting.

Soil Mixes

Although most plants aren't fussy about soil, some plants require specific soil mixes.

Get the most from your indoor garden by cultivating healthy growing habits and promoting successful plants. Potting mix or soil helps provide plants with the water, nutrients, and air they need to thrive. It also anchors a plant in its container.

Indoor plant soil does not come from the yard. Garden soil is too heavy, compacts easily, drains poorly in pots, and may harbor insects or diseases.

Potting mixes

For most container gardens, choose a potting mix. Packaged ready-made mixes usually contain peat moss, coir (coconut husk), or decomposed bark and vermiculite or perlite—no soil—in various proportions. For a potting mix with other ingredients, a wide range of packaged mixes adequate for most plants is available, or you can blend a mix.

Soilless mixes drain well and dry out quickly. Sometimes they contain only peat moss or coir and perlite and are too lightweight to adequately anchor a plant, especially a large or top-heavy plant. Soilless mixes also contain few if any nutrients and require fertilizer. A soil-based mix made with sterilized soil or compost is heavier and holds moisture and nutrients longer. Experiment with both types of mixes and see which works best for your plants.

Custom mixes

Customize a growing mix according to individual plant needs and save money by adding ingredients to standard potting mixes. Potting mixes for flowering plants typically contain more organic materials that retain moisture, such as leaf mold (decomposed leaves) or compost. Moisture-holding mixes usually include water-retentive polymer crystals and suit plants that prefer damp soil. Cacti and succulents need a mix that contains sand and that drains extremely well.

Premium mixes that contain slow-release fertilizer sustain long-term plantings and don't require added plant food for the first growing season.

SPECIALTY BLENDS

Flowering Plants
Mix equal parts sand, peat moss, sterilized topsoil, and leaf mold. Sometimes a comparable blend is sold as African violet mix.

Epiphytes, Orchids, and Bromeliads
Mix equal parts sphagnum moss, coarse bark, and coarse perlite. Add 1 tablespoon dolomitic lime and 1 cup horticultural charcoal to 3 quarts mix.

INDIVIDUAL INGREDIENTS

Perlite
These heat-expanded granules of volcanic ash do not absorb water, but they help potting mix drain and resist compaction.

Vermiculite
Flakes of mica (a mineral) expanded by heat absorb water, release it slowly, and make potting mix more porous.

Choose a potting mix based on the plants you intend to grow. The soil mix should be blended to provide the water, air, and nutrients the plants need.

Cacti and Succulents
A desert-like potting mix that contains equal parts sand, perlite, and potting soil provides porous, fast-draining conditions.

Ferns
Blend 3 parts soilless potting mix with 3 parts leaf mold and 1 part compost. Then add 1 cup horticultural charcoal to 2 quarts of this mixture.

All-Purpose Mix
Combine 3 parts peat moss or coir, 2 parts compost, 2 parts perlite or vermiculite, and 1 part coarse sand. Add slow-release fertilizer and water-holding granules, as desired.

CUSTOM BLEND
Soilless potting mixes work for all kinds of plants and containers and can be customized according to the pot as well as the plant.

Components of potting mixes vary and affect the way the mix works, in terms of holding water, draining excess moisture, and more.

Peat moss
This partially decomposed plant material soaks up water and nutrients like a sponge. Coir substitutes for this limited resource.

Coarse Sand
Tiny rock particles help open up soil and allow air to penetrate. Choose washed or sterilized sand.

Charcoal
Horticultural-grade charcoal absorbs salts and by-products (including odors) from plant decay.

Carefully scratch dry fertilizer granules or pellets into the soil. Water after adding dry plant food to the soil.

Fertilizing Plants

Keep indoor plants looking their best by feeding them with reliable and easy-to-use products.

High-performance plants need nutrients to produce vigorous foliage and bright blooms. Regular fertilizing during the growing season is important to keep potted plants healthy, because frequent watering flushes important nutrients out of the soil. More frequent fertilizing is needed by some plants, including vigorous growers, flowering plants, and those growing in soilless mixes. Less frequent feeding is required by many foliage-type houseplants, citrus, cacti and succulents, and plants that rest or go dormant over winter.

The numbers

Although fertilizer is also called plant food, plants make their own food via photosynthesis. Fertilizer amends the soil with the nutrients that plants need most. Fertilizer labels indicate the percentage of the three key nutrients as a series of numbers, such as 15-30-15, ideal for flowering plants, which means that 15 percent of the product weight is nitrogen; 30 percent is phosphorus; and 15 percent is potassium. Nitrogen is needed for stem growth and green leaves; phosphorus fuels strong roots and flowering; potassium enhances stem strength, disease resistance, and formation of flowers and fruit. An all-purpose 20-20-20 blend suits most plants.

For good health, plants need other minerals (sulfur, calcium, magnesium) and micronutrients (copper, zinc, and others) available in balanced products and noted on labels. When plants are nutrient-deficient, they may show symptoms, such as yellowish or reddish foliage and stunted growth. If you suspect deficiency, apply a soluble fertilizer that is rich in micronutrients.

Types of fertilizer

Plant foods in various forms are applied using specific methods. What matters most is that fertilizer must be dissolved to be of use to plants; soil moisture dissolves nutrients so they can be absorbed by plant roots. Water-soluble crystalline, granular, and liquid fertilizers are convenient to use by diluting them

MULCHING INDOOR PLANTS

Mulch is a layer of loose material that covers the surface of the potting mix in a container and works primarily to conserve soil moisture. It also prevents soil from splashing on foliage and washing out of a pot when it is watered. If you move plants outdoors in summer, mulch insulates soil and plant roots, keeping them cooler during hot days. Mulch can also deter cats and squirrels from scratching in potted plants. It can be a colorful and pretty embellishment. Apply a 1-inch layer of mulch—polished stones, glass beads, crushed recycled glass, marbles, shells, rinsed gravel, or moss—evenly over soil.

with water. A solution reduces the potential for burning caused when excess fertilizer damages roots and leaves. Dry plant foods include granules, powdery crystals, slow-release pellets, and spikes. You can make multiple applications of a soluble fertilizer that is taken up relatively quickly by plants and used in about a month, or you can make fewer applications of a less-soluble fertilizer. Slow-release formulas are made for a single application each growing season and last for six to nine months under average conditions. Organic fertilizers include compost, fish emulsion, worm castings, and kelp products. Edible plants and long-term plantings benefit from organic plant foods that enrich soil and improve its structure.

Applying plant food

Follow package directions when using fertilizer. Fertilizing is not advised when plants are wilted, sickly, or lacking adequate light. Under most circumstances, fertilize plants regularly during spring and summer, when they're actively growing. Skipping or missing a dose of plant food does not have the same deleterious effects as forgetting to water. Overfertilizing hurts plants more than it helps them and shows up as burned leaf edges, poorly shaped leaves, and white crust on the soil surface or inside the pot rim. Occasionally leaching or rinsing soil minimizes salts that build up from fertilizer and water. If a plant is sensitive to the mineral salts in fertilizer or the fluoride in water, replace its soil periodically.

Grooming Plants

Remove dead leaves and prune excessive growth for healthier plants.

All plants share particular preferences for moisture and heat, depending on their origin. But most plants adjust to indoor conditions, especially with the gardener's awareness of needs and assistance in fine-tuning environmental variables. As conditions change seasonally, pay attention to how indoor plants are affected.

From tidying to pruning

The first step in grooming is to remove old and dead leaves as well as faded flowers. Some withered leaves fall off on their own and should be tidied up to eliminate harbors for pests and disease. Yellow or brown leaves won't turn green again. Pluck or snip them off to make the plant more attractive. If large leaves have brown tips or edges, trim off the brown section using sharp scissors and let the remainder of the leaf recover. Flowers that aren't destined to become fruit, such as those on African violets or hibiscus, should also be plucked when spent. Removing dead flowers often makes way for new ones.

Pinching back and pruning improves plant shape and promotes healthy new growth. Pinch or snip off the growing ends of fast-growing or untidy plants that show signs of unruliness or unattractive bare stems. Cut or pinch just above a node where new leaves or branches can develop. Although some plants may never need pruning, others require it to encourage and direct growth; to remove damaged, diseased, or dead branches; or to adjust plant height or width. Keep the natural shape of the plant in mind when pruning. If you're apprehensive about cutting off a branch, cover it with your hand before you make the cut, and picture how the plant will look without the branch.

left Bromeliads thrive with regular misting as their source of water. Although misting does little to boost humidity, it can rinse dust off leaves. Dusty leaves absorb less light. ***opposite, below left*** Remove yellow leaves from plants to ensure health and make them look neater.

Using your fingertips to pinch out the stem tip and its topmost leaves promotes side branching and bushier growth. Coleus, nerve plant, and peperomia are among many plants that benefit from frequent pinching. Trailing plants, such as Swedish ivy and pothos, also respond positively to periodic pinching.

Dusting plants

Use a household duster or soft cloth to remove dust from leaves at least once a month. Dust particles block light and clog leaf pores, interfering with transpiration. Hold one hand under a leaf, supporting it as you dust with your other hand.

Another way to remove dust is to bathe a plant. Place small potted plants in a sink and spray them with tepid water, using a spray nozzle. Place large plants in the shower. To prevent soil from washing out of the pot, slip the container inside a plastic bag, pull the top of the bag up over the soil, and cinch it around the plant stem. Secure the cinched bag with twist ties or twine.

HOW TO GROOM PLANTS

Regularly clean up any litter—fallen leaves, flowers, or twigs—that collects on soil and between potted plants. Also, when watering plants, rotate pots one-quarter turn clockwise to promote even growth. Other practices that bolster plant health include repotting, adding fresh soil to pots, fertilizing, and mulching.

DUSTING Use a soft, damp cloth to wipe dust off large glossy or waxy leaves. As you work, frequently rinse the cloth in clean warm water.

PRUNING Always use clean sharp pruners to trim plants. It's best to prune during spring or summer, when a plant will grow quickly to cover the pruning wound.

PINCHING Pinch or cut stem tips of plants such as aluminum plant that have soft stems. Pinching prompts plants to send out new shoots, making the plant full and lush.

BRUSHING Dust plants that have fuzzy leaves (African violet, purple passion plant, some begonias) with a soft bristle brush, such as a cosmetic brush or paintbrush.

above Pothos, philodendron, spider plant, and others happily move outdoors to spend the summer in a shaded spot. They create a pretty display outdoors too. ***below left*** Before moving a plant indoors after a summer vacation on the patio, give it a strong shower, using a garden hose to blast off any hitchhiking insects. Rinse the tops and bottoms of leaves. ***left*** A two-wheel dolly makes it easy to lift and haul large potted plants indoors and out. Large pots set on caddies with casters can become mobile.

Plant Vacation Care

Move houseplants outdoors for the summer to reduce indoor watering chores while giving plants a natural break.

Many indoor plants enjoy a summer respite outdoors in mild weather as much as gardeners do, especially after a long, dark winter. Think of it as giving your plants a summer vacation. They serve in the process by adding tropical appeal to outdoor living areas.

Outdoors, plants will thrive with filtered light, increased air circulation, and refreshing rain showers. Because plant health and appearance improves as energy recharges, you might notice more lush foliage, saturated color, vigorous growth, and profuse flowering. Natural insect predators can also go to work to rid plants of pesky mites and scale.

Outside for the summer

Take outdoors only the toughest plants—those that can handle periodic wind, hard rain, and unexpected chill. Wait until early summer, when warmer weather patterns have settled in, then gradually move plants outside. Help them adjust to the change in environment by setting them outside for a couple hours one day, and a few hours the next, increasing time outside each day. Place them in a protected place with afternoon shade to prevent damage from wind, hail, and sun. Some houseplants, such as cacti, tolerate full sun outdoors once they are acclimated to the summer environment. Keep low-light plants under trees, shrubs, or a structure in full shade. Monitor soil moisture carefully— potted plants dry out much faster outdoors, especially when weather is warm and breezy.

Moving back indoors

Bring plants indoors in late summer, before temperatures drop at night. If you wait until early fall to make the transition, plants will have to adjust to changes in temperature and humidity in addition to lower light level, which is stressful. Ease plants through the adjustment again by moving them to a shadier place outdoors for a couple weeks before moving them indoors. Inspect and groom all plants to minimize any chance of bringing pests and diseases indoors with them. Consider repotting any plants that have grown substantially over the summer—completing this task outdoors leaves the mess outside. If many plants have vacationed outdoors, bring in a few at a time, giving each one careful attention. Delaying one big move until the last minute when frost threatens stresses gardeners and plants.

While you're away

Plants should fare well for a week if you leave them well-watered and out of direct bright light, to prevent overheating and sunburn damage. If you plan to be away two or three weeks, consider these options: A thorough watering, followed by tenting each plant in a spacious plastic bag will keep most plants supplied with adequate moisture for several weeks. Leave some openings for air circulation and insert stakes into the pot to prevent the plastic from touching the leaves. Alternatively, use a wick-watering system. Place a large container of water next to the plant and extend narrow tubing or an absorbent shoelace from the water into the soil. As the soil dries, water wicks into it via capillary action. Let the lower end of the tubing or shoelace reach down to the bottom of the water container.

BHG TEST GARDEN TIP

PRETTY COMBOS

Keep plants in their usual pots when you move them outdoors or combine plants with similar light needs in spacious containers such as self-watering buckets.

Pest Control

Indoor plants can be bugged by a number of pests. Keeping them healthy is the best defense.

An ounce of prevention truly pays off when it comes to plant pests, and it is doubly important within the confines of an indoor garden. Healthy plants are amazingly able to resist problems, but rascal insects can ride in on a new plant or their eggs can hide in potting soil. If you neglect your plants, forgetting to water and feed them, the resulting stress will make them more susceptible to insect infestations and diseases.

Minimizing pest problems
Make it part of your routine to check for signs of problems whenever you water plants because you can correct a problem more easily if you catch it early. Look for obvious signs (insects, eggs, sticky sap, webs) and subtle clues (weak, pale, or distorted growth; yellow or dropped leaves; stippling or tiny yellow dots on leaves).

When you notice a change—the Boston fern is shedding more leaves than usual, the hibiscus has more than a couple yellow leaves, one or two fungus gnat sightings a month have turned into daily appearances—take a closer look. Problems like these need to be tended before they spiral beyond control.

Identify the problem
If you notice wilted leaves on a plant, consider the possibilities: Does the soil feel dry? Is the plant getting too much sun? Has it outgrown its pot? Are there any signs that indicate disease? Distinguishing between a cultural problem and a pest or disease is half the solution.

above left Before you take steps to remedy a pest or disease, identify it correctly. A magnifying glass helps a lot. Here, the problem appears to be a mealybug. *below left* Check the backs of leaves for insect pests before you bring plants indoors for the winter.

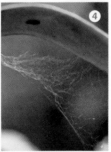

Resolving problems

In time you'll recognize the difference between a healthy plant and one that needs help. If you spot signs of an insect or disease problem, move the affected plant away from others while you take steps to solve the problem. Choose the mildest remedies first: Pick off insects or affected leaves, then rinse the plant with a solution of soapy water (using Fels Naptha or castile soap; most dish soaps are too harsh for plants). If the problem persists, apply horticultural oil or insecticidal soap. Repot the plant if it has a root disease or soil pests. If the problem continues, prune the affected parts. If you use an insect- or disease-control chemical formulated for indoor use, check the label to determine if it is also formulated to treat the specific problem affecting your plant, then follow directions. The last resort: Discard the plant.

After dealing with the problem, reevaluate the plant's growing conditions and adjust the light, water, humidity, temperature, soil, or fertilizer accordingly. Have patience—the least toxic methods of pest control may take some weeks to resolve the situation. In the meantime, focus on bolstering the plant's health to aid its recovery.

BHG TEST GARDEN TIP
STICKY TRAP

Cut a 6×8-inch piece of yellow cardstock. Brush petroleum jelly or Tanglefoot (from a hardware store or garden center) on both sides of the card. Hang it near plants to trap insects.

COMMON INDOOR INSECT PESTS

Damage to indoor plants is common because natural predators rarely reside indoors. Become familiar with the most common pests and employ simple methods to eliminate them.

1. MEALYBUGS These cottony insects suck plant sap from leaves and stems. The plant weakens and new growth is deformed. Wash the plant with soapy water; apply rubbing alcohol to the mealybugs. Use neem or horticultural oil.

2. SCALE Crusty brown or reddish-gray bumps or cottony white masses appear on stems and leaves. Leaves turn yellow, drop, or collect sticky residue. Repeated applications of horticultural oil smother adult insects.

3. FUNGUS GNATS Dark fungus gnats fly around plants and crawl on soil, leaves, and nearby windows. They lay eggs in soil; larvae feed on roots and kill seedlings. Use a yellow sticky trap to snag adults; use Bt (Bacillus thurengiensis, a naturally occurring bacterium) to kill larvae.

4. SPIDER MITES These troublesome pests thrive in heat and low humidity. Look for stippling on leaves, webbing in extreme cases. Adjust conditions; bathe the plant. Use horticultural oil if necessary.

5. WHITEFLIES Feeding mainly on the undersides of leaves, the tiny insects create a sticky residue. Use yellow sticky traps. Resort to insecticidal soap for extreme cases. Air circulation is vital for prevention.

Beating Disease

Plants can become stressed, which causes a variety of ailments. Symptoms will indicate treatment.

Disease is seldom found on a thriving plant. Ensuring that plants get what they need in terms of soil, light, water, and food is critical. Sometimes problems occur due to an issue with one of these elements—too much or too little, whatever the case may be. Then, a weak, damaged, or stressed plant is more prone to disease.

Bacterial diseases, although uncommon, proliferate by rapidly reproducing and damaging plant tissue. A fungal disease takes hold when a spore contacts a leaf, becomes a parasite, and destroys plant tissue. A virus, often present but dormant, in certain plants, infects a stressed plant and weakens it. Viruses are carried from plant to plant by insects or people, or on infected tools.

Disease prevention

Regular grooming prevents many problems. Immediately getting rid of leaves or stems that show signs of trouble is good practice, regardless of the cause. Allowing fallen dead leaves to remain on the soil can exacerbate a problem. If one of those leaves has a disease, even in the early stages, the disease will be nurtured and spread via damp soil. It's easy enough to pick up dropped leaves and put them in the trash, which may totally prevent or minimize the problem.

The presence of disease usually indicates an issue with plant culture or conditions. Powdery mildew, a fungus that appears as white powder on leaves, is an indication to improve air circulation, to lower humidity and temperature, to decrease watering, and to prune out diseased leaves. By correcting these cultural issues, you'll fend off other, more serious diseases.

left A healthy plant has natural vigor and shine. Bromeliads, such as this Neoregelia carolinae 'Franca', tend to be free of disease, but unfavorable growing conditions can cause problems. **opposite, bottom left** Clip off the dead or damaged parts of leaves to help keep plants healthier and to improve appearance.

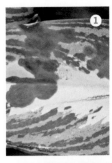

Typically more harmful and difficult to control than pests, diseases develop in a greenhouse or at home. Before you buy a plant, look for signs of problems, especially viruses, which kill the plant they attack and spread easily to other plants. Isolate any new plants for a week or two after bringing them home to make sure they are free of diseases and insects. Understand the plants' needs and give them the proper care they need to reduce stress and prevent problems.

Problems and solutions

When plants show signs of trouble, a single symptom can have different causes. Yellow spots on leaves may indicate one or more problems, including bacterial infection, viral infection, or sunburn, depending on the size, shape, and prevalence of the spots. Learning to identify diseases can be tricky. Sending a photo or plant sample of your plant to a nearby university extension plant clinic will be helpful.

COMMON INDOOR PLANT DISEASES

Once you have identified a plant's problem, choose the appropriate solutions. The top line of defense should be to learn about and provide the best conditions for the plant.

1. BACTERIAL BLIGHT Soft, sunken areas with water-soaked margins appear on leaves or stems. Leaves may turn yellow and severely wilt. Inner stem tissue turns brownish. Take a healthy cutting or two and discard the plant. The condition is not curable.

2. LEAF SPOT Caused by bacteria or other pathogens, the disease is seen on leaves as circular reddish-brown spots with yellow halos that may join to form blotches. Snip off badly spotted leaves. Water carefully to keep leaves dry.

3. POWDERY MILDEW White or gray powdery fungal patches form on leaves, stems, and flowers. Remove infected parts. Give plants adequate light and air circulation. Use potassium bicarbonate to prevent infection.

4. ROT Excessively wet soil stresses a plant. Fungi proliferate and affect the leaves, stems, crown, or roots. The affected part appears dull and turns brown or black and mushy. Let the soil dry between waterings. Remove and destroy damaged parts.

5. VIRUS The plant grows slowly, without vigor. Leaves are mottled or streaked with yellow and distorted. No cure exists. Infected plants should be discarded. When pruning, dip tools in a disinfectant (diluted bleach or rubbing alcohol) between plants.

CHAPTER FOUR

Plant Propagation

Learn various propagation techniques to get the most from indoor plants by dividing and starting new plants.

HOW TO MAKE STEM CUTTINGS

Prepare nursery containers by filling small pots or cell packs with premoistened soilless potting medium. Many potting soils hold too much moisture.

1 CUT
Use a clean, sharp knife to cut a 3- to 4-inch-long nonflowering stem. Cut off the bottom leaves.

2 DIP
To encourage root growth, dip the cut end into rooting hormone powder.

3 PLANT
Poke a planting hole in the damp medium, using a chopstick or pencil. Insert the cutting.

4 HUMIDIFY
Place the pot in a plastic bag held closed with a twist tie to create a humid, greenhouselike environment.

5 TRANSPLANT
Set cuttings in bright, indirect light. Water as needed to keep the medium damp. When roots develop, transplant into a larger pot.

Cutting Techniques

Creating more plants from cuttings is an inexpensive and fun way to multiply houseplants and to share the wealth with friends and family.

Add to your plant collection without subtracting from your bank account by learning to take cuttings. Growing houseplants from cuttings is the most popular form of vegetative propagation because it is so easy and rewarding. An array of plants, from annuals to perennials and shrubs, will grow well from cuttings and be ready to transplant within a month or two. Just be sure not to take cuttings of trademarked or patent-protected plants that state "propagation prohibited by law" on the tag.

Stem cuttings

The most common propagation methods include taking cuttings of young, green (softwood) stems from late spring until midsummer when plants are growing strong, or from partially mature, current-season (semi ripe) stems from June through August. Depending on the plant, you can take cuttings from stems as well as leaves or roots.

Cuttings will root in various media: soilless potting mix, vermiculite, commercial rooting gel, or water. Although it doesn't work for every plant, rooting begonia, Swedish ivy, and purple passion plant in water is super easy. Just cut a 3- to 4-inch-long stem tip, snip off the lowest leaves, and tuck the cutting in a jar of water set in bright, indirect light. Change the water weekly to keep it fresh and to minimize bacteria growth. When roots develop, transplant the cuttings into a container filled with potting mix. The roots that form in water are more coarse and fragile than roots formed in soil, so keep the potting mix evenly damp for the first couple of weeks to help the plant adjust and develop terrestrial roots.

Leaf cuttings

Take leaf cuttings of plants with thick leaves or leaf stems, such as African violet, begonia, peperomia, and succulents of all kinds. Remove the leaf as close to the parent plant as possible. Dip the cut end in powdered rooting hormone and tuck it into damp rooting medium, following the directions for stem cuttings. After a few weeks, roots will start to form; within a few months, a new shoot with leaves will emerge.

Root cuttings

To start a new plant from a root cutting, cut a 2-inch-long section of thick root and place it horizontally in rooting medium. When a new plant develops, transplant it as you would other cuttings that have rooted. Plants suitable for root cuttings include ti plant and arrowhead plant.

BHG TEST GARDEN TIP

ROOTING MIX

Make a propagating medium that works well for leaf cuttings (right), stems, or roots: Mix equal parts sterilized coarse sand, perlite, peat moss, and vermiculite.

opposite, above Cuttings of plants that root well in water include tradescantia, scented geranium, coleus, begonia, and ivy.

Plant Division

Not only does plant division make additional plants, but this process is also necessary for the health of some plants.

The immediate result of division is two or more plants of respectable size. When dividing a plant, pull apart clumps that have multiple stems arising from the soil surface and separate underground roots for each clump. Then pot each new section or plant division in separate pots.

Candidates for division

Division is an excellent way to rejuvenate a plant that has outgrown its pot and its setting—the plant appears top-heavy or roots have filled the pot. This is an ideal opportunity to share a favorite plant with a gardening friend, rather than repot the plant into a larger pot. Spring is the ideal time to divide plants, when they are on the verge of the growing season.

Plants commonly propagated by division include Chinese evergreen, pothos, snake plant, (how-to shown at right), Boston fern, asparagus fern, peperomia, and cast-iron plant.

Dividing tubers and rhizomes

Caladium and tuberous begonia are among the plants that produce plump underground growths called tubers. As long as roots and growing points are present on the tubers, they can be separated and established as new plants. Alocasia and zeezee plant are among the plants that grow from tuberous rhizomes, underground stems of sorts that produce shoots above and roots below.

Divide tubers and rhizomes in fall, when the plant is not flowering and is beginning a period of rest or slower growth. To divide a rhizome, cut it into pieces, making sure that each piece has a growing point. Then plant each piece individually. Over time, tubers often develop two or more distinct growing points. At planting time, just before potting, carefully cut between two growing points to yield two separate tubers.

opposite When dividing a plant into new clumps, sometimes tease apart the roots by hand and at other times cut them with a sharp knife or pruning saw.

HOW TO DIVIDE A PLANT

Easily propagate one plant into two or more. Division consists merely of separating a plant into sections, then potting each section individually.

1 DIVIDE
Remove the plant from its pot. Separate the leaves into clumps and pull or cut apart the root ball into corresponding clumps. You may need to use a pruning saw or large serrated knife to saw some plants apart; others break apart by hand. Make sure each division includes some of the main root and stem system.

2 SECTION
Set up a new pot with fresh potting mix for each division. Position the division in the center of the container, giving roots room to expand. Gently set the division into the pot.

3 REFRESH
Plant each clump in its new pot, with the base of the plant at the same level as the original planting. Add potting mix around the stems, taking care to cover the roots.

4 WATER
Water the new planting, thoroughly moistening the potting mix and settling the new planting into it. Add more mix as needed.

above top Use a clean sharp knife to cut off offsets. A clean cut ensures a healthy plant. ***above bottom*** Piggyback plant (Tolmeia menziesii) makes new plantlets at the base of its leaves. These plantlets root easily when placed in soilless potting mix. ***left*** Most homes, even small apartments, offer a variety of light conditions. Each setting is an opportunity for plants. Foliage plants usually need less light than flowering ones.

Plantlets and Offsets

Can plants have babies? Well, sort of. Some plants develop miniature versions of themselves that can be potted up.

Two common methods of propagating plants are especially easy because the parent plant does most of the work by producing plantlets or offsets. In both cases, the parent plant forms miniature replicas of itself.

What are plantlets?

A plantlet develops at the end of a long stem or runner—a lifeline—or on a mature leaf. A plantlet develops roots while it is tethered to the parent plant. The long, flexible stems of spider plant and strawberry begonia that connect the older and younger plants can be bent enough for the plantlet to be nestled into soil in the pot of the parent plant or in its own small pot. Within a few weeks, the young plant has enough strong roots to survive on its own. Cut the tether and the baby becomes a fully independent plant. Species that produce plantlets on their leaves include piggyback plant and a few kalanchoes. The plantlets of piggyback plant can be snuggled into damp soil and secured with a hairpin or U-shape wire. The plantlet will soon root and become a full-fledged plant.

The plantlets of some houseplants, such as spider plant and tradescantia, start to develop roots before they contact soil. Encouraged to root more in damp soil or water, the plantlets quickly comply and can soon be separated from the parent plant.

What are offsets?

Offsets form at the base of a parent plant and remain attached to it. When an offset is big enough to survive on its own and—ideally—has its own roots, it can be separated from the parent plant and planted in its own pot. Common plants that propagate themselves with offsets include many bromeliads, cacti, and succulents.

HOW TO PERPETUATE PLANTLETS AND OFFSETS

Growing new plants is easy when you imitate nature and follow its lead. These small replicas of the parent plant—plantlets and offsets—root easily.

PLANTLETS Strawberry begonia produces plenty of identical offspring, or plantlets, ready for rooting in a nearby pot. Secure a plantlet in damp potting mix, using a hairpin or bent wire. When the plant roots, sever the connection to the parent plant.

OFFSETS Peanut cactus forms offsets at the base of the parent plant. Snap off or cut away each offset, with or without roots. Snuggle it into damp potting mix, keeping it damp until roots form. It should grow easily.

Layering Techniques

This method of propagation is similar to rooting cuttings, except the portion to be rooted remains on the parent plant.

In many cases where cuttings will not work, layering can be used to make new plants. It is also a good way to rejuvenate a plant that has become overly leggy or lost its lower leaves and become top-heavy. The technique of layering is similar to rooting cuttings, except that the part of the plant (usually a branch or stem) to be rooted remains attached to the parent plant while rooting is in progress.

The advantage of layering is that the parent plant supplies the new plant with water and nutrients while its roots form. Daily maintenance is unnecessary. For slow-to-root plants this is a distinct advantage. At the same time, there is a disadvantage: New plants formed by layering develop roots more slowly than cuttings. The process takes a few weeks to several months before the new plant is ready to be separated from the parent plant and potted in another container.

Layering techniques

Layering can be accomplished in soil or air. Creeping and trailing plants, such as pothos and philodendron, are ideal for layering in soil. Air layering proves most suitable for plants with woody or thick stems, such as diffenbachia, croton, dracaena, rubber tree (*Ficus elastica*), fiddleleaf fig (*Ficus lyrata*), and ti plant.

Soil layering

A suitable plant for soil layering has one or more stems low enough to bend it down to touch soil. Some plants self-layer, developing roots where stems touch soil, in which case, the newly rooted plant can be snipped and transplanted. Otherwise, layer by setting a pot full of damp potting mix next to the parent plant, then anchor a trailing stem in the soil mix while leaving the growing tip free. Layer more than one stem from the parent plant in the pot as long as each stem contacts the soil mix. When the layered stems have rooted, cut them free from the parent plant.

opposite This ti plant presents multiple opportunities for using the air-layering technique to propagate several lush new plants from a gangly mature one.

HOW TO AIR-LAYER A PLANT

A ti plant is one houseplant that benefits from air layering. As the plant matures, it develops tall stems without lower leaves. The new plant will be shorter and more lush.

1 NOTCH
Holding the stem securely, make a slanting cut upward, no more than halfway into the stem. Cut again to remove a sliver of the stem. Continue holding the upper stem to prevent it from falling over and snapping off.

2 WRAP
Wrap a handful of moistened sphagnum moss around the layering site. Encase the moss with plastic wrap. Secure the top and bottom edges of the wrap with electrical tape or twist ties.

3 ROOT
Allow roots to fully develop and begin to fill the encasement, which often takes several months. Open the top of the wrap weekly to add water to the sphagnum moss, keeping it moist.

4 DETACH
When the roots are well-developed, remove the tape and plastic. Cut the stem below the roots. Transplant the new plant into a pot of fresh potting mix. Water thoroughly.

Starting Plants from Seed

Raising houseplants from seed takes patience, but for the avid gardener, this is an exercise in creating life.

Most houseplants are propagated from cuttings or divisions, but some can be grown from seeds or spores. Seed starting provides the most economical route to raising a wide variety of annuals and edibles. Ferns and bromeliads have many more family members than the market offers widely, but with patience, you can grow rare ones from seeds or spores.

When to plant

Seed starting is a wonderful late-winter and early-spring activity for indoor gardeners. Before the outdoor gardening season gets into full swing, you can coax seeds into seedlings within weeks. Gather a large bag of soilless seed-starting mix, plastic flats with inserts (or any kind of clear plastic container), plant labels, and a waterproof marker. A winter greenhouse can be as simple as an arrangement of shelves with a fluorescent shop light hung from each shelf or a roomy table with a shop light propped above a flat or two.

Most seeds are sold in seed packets that include detailed information about sowing and raising the plants, from how deep to plant the seeds to how many weeks the seedlings need to grow before they're ready for transplanting.

left Make as many free seed-starting pots as you can use, rolling 4-inch-wide strips of newsprint around a 2-inch-diameter glass. Fold under the bottom edges to shape a biodegradable pot. ***opposite, below left*** Polka dot plant seeds germinate in about two weeks.

Caring for seedlings

Seeds have varying requirements for germination (sprouting). Some seeds need warmth; others need a chilling period in order to germinate. Some seeds require light and should not be covered with seed-starting mix. Others need to be covered. Push them into the soilless medium with a pencil eraser. See "How to Start Seeds," at right.

When seedlings emerge, remove the cover from the germination container, and set the tray under a fluorescent light equipped with one cool- and one warm-spectrum fluorescent bulb. Keep the soilless mix damp by watering from the bottom of the container. By the time seedlings are big enough to transplant—with at least two sets of leaves and roots that have begun to fill the cell pack—they're ready for a dose of half-strength fertilizer.

HOW TO START SEEDS

To sprout houseplant seeds, a few inexpensive supplies are needed. Research the species you want to grow, become acquainted with basic needs of seeds, and grow your own plants.

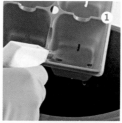

1 CLEAN
Start with clean seed-starting containers that allow excess moisture to drain away.

2 MOISTEN AND FILL
Moisten soilless seed-starting mix with warm water then fill the cell pack or other container with the mix.

3 SOW
Sow one or two large seeds per cell; sprinkle three or more small seeds into a cell. Follow seed packet instructions for sowing depth, and cover them with soil accordingly.

4 WATER
Sprinkle soil with water and cover the container with clear plastic to hold in moisture.

5 WARM
Warming the soilless medium using an electric heating mat beneath the tray can speed up seed germination.

CHAPTER FIVE

Creative Planting Projects

Gather ideas, inspiration, and tips to fashion all sorts of indoor gardens any time of the year.

left Succulent kalanchoe blooms for weeks on a sunny windowsill. Removable glass shelves span the window and let plenty of bright light reach the plants. ***above, top*** A metal shelf mounted on a window frame keeps herbs within easy reach for cooking. Herbs and other edibles do best in the sunniest south- or east-facing windows. ***above, bottom*** The direct, bright light of a south-facing window sustains a collection of cacti and succulents. The individually potted plants live contentedly in a drywaller's pan. Gravel hides the pots.

Windowsill Gardens

Windows and indoor plants were made for each other! Indoor window boxes filled with light-loving plants will improve the view from inside or out.

A few cheery plants on a sunny windowsill bring life to any home and add a touch of warmth on drab days, especially in areas where winters are cold. If you have only one sunny window in your home, use it for a garden, whether you arrange small pots on rows of shelves across the window or extend the sill by bumping a tabletop up to it. Use a window to frame a plant display, making it more artful. And use plants as living curtains to obscure the view into your home.

Light requirements

Providing plants with enough light indoors is crucial. How much light is enough? That depends on each plant and the conditions in your home. You can find plants to suit almost any indoor environment. Most plants will do fine with slightly more or less light than optimal, especially for a limited time.

Widely varying levels of light are needed for plants to grow, and the levels of light from windows vary too. One type of plant may simply survive while another fares well with low light. In some cases, the plant may grow well, but won't bloom in the low-light conditions. Flowering plants typically need high light from a south- or west-facing window. If there is enough sunlight in a room to read or do handiwork without turning on a light, it will be a suitable place for plants—such as a range of houseplants that need low to medium light. Given enough bright light, traditional outdoor plants, such as culinary herbs or salad greens, can grow indoors.

Increasing light levels

Most indoor settings are actually darker than they seem to the human eye. If it is difficult to read a newspaper in your home at midday, supplemental light is needed. Similarly, unless plants that need bright light are situated directly in front of windows, they will need added light. Plants show signs of inadequate light when they fail to grow well, have lighter green or smaller leaves than normal, or become spindly and stretch toward a window or other light source.

Strategies for increasing the light that plants receive include keeping windows clean, removing screens when they're not needed, hanging mirrors strategically to reflect light, painting walls flat white or light colors to reflect light, periodically washing or wiping foliage, and replacing small windows with larger ones.

CAN-DO SHELVING

Choose from an array of stylish-yet-sturdy shelf brackets made to hold glass shelves. Easily mount each bracket on a window frame using a couple screws. Get cut-to-fit shelves in a desired width and thickness from a glass, mirror, or windshield supplier. Save money by comparing prices before ordering.

Dish Gardens

Tabletop dish gardens are ideal for growing a variety of foliage, flowering, and succulent plants.

A dish garden presents an opportunity to grow a group of plants in a single container on a tabletop or a windowsill. Requirements are few: A shallow container and a group of shallow-rooted plants, such as cacti, succulents, or small houseplants, comprise a dish garden. A small, low-growing design for a tabletop proves engaging and won't interfere with across-the-table conversations. A large dish garden can be quite dramatic and colorful.

Making a dish garden

As you decide on a container for a dish garden, consider the conventional—a deep terra-cotta saucer or a shallow ceramic bowl—or be more creative. Transform a cup and saucer, vintage roasting pan, or hubcap into a dish garden. If the container has a drainage hole, the dish garden will need less maintenance. A container without built-in drainage must be watered very carefully to avoid waterlogging plant roots, unless the dish garden is made with water-loving plants, such as dwarf papyrus (Cyperus) and fiber optic grass (Isolepis).

Prepare the container for planting by covering the bottom with a thin layer of horticultural charcoal, which helps prevent soil from becoming funky when it holds too much moisture and not enough air. Top with a 2-inch layer or more of potting mix.

Group plants with similar soil, moisture, and light needs. A classic dish garden that features succulents in well-draining soil and placed by a warm sunny window can go a couple weeks without watering. Dish gardens are low-maintenance. A leafy plant grouping benefits from the higher-humidity microclimate that results from proximity to each other. Massing several of the same begonia or similar plants with showy foliage intensifies their effect too. Choosing slow-growing plants helps prevent overcrowding.

Planting arrangements

Before you plant a dish garden, keep plants in the pots while playing with planting schemes. Try the tallest plant in the center of the arrangement with the shortest plants at the perimeter. Or try an asymmetrical grouping, with the tallest plant off-center and a cluster of smaller plants balancing the display. Leave growing room between plants. When not overcrowded, it is easier to remove an ailing plant without disrupting the roots of other plants. Trimming plants periodically can also prevent overcrowding.

Some dish gardens benefit from an accent or focal point. Naturally fitting objects include driftwood, stones, or seashells. A decorative figurine may be the best way to give a dish garden your signature touch. Covering the soil surface with gravel can be a decorative and functional finishing detail.

opposite, left This hot combo of houseplants includes 'Red Hot' flamingo flower, a rex begonia with silver streaks, and purple passion plant in a cool blue bowl. *opposite, top right* A deep terra-cotta saucer provides a good home for succulent kalanchoe as long as it also contains ample pea gravel and sand for drainage. *opposite, center right* Colored gravel or tumbled glass adds contrast to a group of succulents and cacti in a contemporary dish garden. *opposite, bottom right* A variety of small succulents contributes to an interesting dish garden that works as a centerpiece and a topic of conversation.

Little Landscapes

Fairy gardens planted with tiny plants and acorn-size furniture are whimsical fun for children and adults.

A miniature garden, typically contained within a box, tray, or dish, presents another realm of gardening. Making a teeny-tiny plantscape complete with wee furnishings, itty-bitty plants, and plenty of hide-and-seek mystique inspires a playful sort of garden design. The closer you approach to view it, the more beautiful and complex the garden appears. What's more, this could be a small-scale version of your dream garden.

Tiny gardens with big appeal

Little landscapes may be even more charming than full-size counterparts. Imagine a garden that requires few materials, novice planting skills, and minimal upkeep. Once planted, a miniature garden can be long-lived, moved outdoors to a partly shaded spot over summer, and then moved indoors to a sunny windowsill during winter.

Miniature gardening proves so engaging and entertaining that it has become a popular hobby for adults and children alike. Creating a garden fit for a fairy is an ideal opportunity to get kids to dig into gardening and stretch their imaginations. Calling it a fairy garden plays up its whimsical nature, an escape to a place where mythical fairies dwell. But you don't have to believe in pixies to be enchanted by these miniature gardens.

Making a mini garden

A well-crafted miniature garden begins with a plan. Establish a theme, such as a pint-size patio or formal knot garden. As you proceed, keep everything to scale for best results.

Design a tiny landscape to live indoors and outdoors, if you like. It's a good project for a rainy day indoors that can bring pleasure for months ahead. During the gardening season, move the container outdoors and give it a special hideaway where it will provide surprise and delight for anyone who discovers it.

Miniature tuteurs, tiny bricks, and pea gravel create the basis for this small-scale formal garden. Asparagus "trees" frame a pergola seating area that overlooks hens-and-chicks (Sempervivum) and air plant (Tillandsia).

Container: Be creative or settle on a rough-and-ready holder that will weather indoor-outdoor life. Containers should give roots room to grow and allow proper drainage to prevent soggy roots. Set the container on a tray to protect surfaces where it is displayed.

Plants: Choose dwarf (slow-growing) varieties that have similar needs for light and water. Plants of diminutive size with small flowers and leaves work best. Dwarf conifers and herbs that thrive in sun with once-a-week watering minimize maintenance. Faster-growing plants require frequent trimming and eventual replacement. Tiny standards or tree-form plants, such as myrtle (*Myrtus*) or euonymus topiaries, are in scale in a miniature setting. Young herbs, such as rosemary or thyme, serve as itty-bitty shrubs that can be trimmed easily into neat forms.

Accessories: Collect objects related to the garden theme, from tiny furniture to fences, pots, and tools. Notice potential objects around your house and yard: A marble becomes a gazing ball; a birdhouse provides a garden shed. These details will make your miniature garden even more fascinating.

TINY TABLETOP GARDENS

It's so easy and fun to create a miniature garden, chances are you'll make more than one. Start with a container and a theme, add a focal point, then tuck in plants, and finish with details.

Vintage Container

A metal picnic box adds to the whimsical quality of a fairy garden. To provide drainage, drill holes in the bottom of the container, or cover the bottom with 1 inch of gravel before adding potting mix.

Scaled Down

A glass tray or similar container for a mini garden keeps the simple scheme. This tidy, formal design includes only one plant, Scotch moss (Sagina subulata). Pea gravel forms pathways, and preserved moss creates beds of green. Different colors of moss and gravel are useful in any mini garden.

Junk Drawer

This pint-size plot demonstrates the thrifty side of miniature gardening. Set in a repurposed drawer, the design uses bits of broken terra-cotta pots to shape a patio. Garden suppliers offer an array of tiny accessories, such as wire chairs and picket fencing.

In the Round

A miniature knot garden in a standard pot achieves a design to view from any angle. Based on historical garden designs with patterned plantings, the hedges of this display are made with grass and moss. The dwarf conifer and urn could be replaced with a little sundial or figurine for variety.

Choose small-scale plants (and equally tiny garden accessories) to create a magical tabletop garden.

PLANTING MADE EASY

Using a wooden skewer or comparable tool works better than fingers to plant tiny root balls and position other elements in the confines of a teeny garden.

Wooden Planters

From vintage crates to newly constructed planters, wooden boxes are ideal containers for indoor gardens.

Containers made of wood, from stylish boxes to sturdy crates, often provide desirable homes for indoor gardens. Wooden containers, easy to find and usually economical, typically have simple lines and suit a range of decorating styles, from raw and rustic to sleek and modern. Part of the beauty of wood is that you can change its appearance to suit your home's style by adding a coat of paint, stain, or other finish.

Protect your investment

When choosing a wooden container for a garden, consider its potential to disintegrate when exposed to wet soil. Some woods, such as cedar and teak, withstand challenging conditions better than others. Use a container that's weather-resistant if you plan to expose it to outdoor life. To preserve the integrity of a wooden container, apply several coats of exterior waterproofing sealant to the interior. Seal the exterior too, or protect it with two coats of exterior-grade paint or stain.

Ready for planting

Paint an inexpensive wooden container from a crafts store for the project shown. Combine a graceful weeping pussy willow (Salix) with a selection of small spring-flowering bulbs and velvety moss for a delightful tabletop garden that lasts for weeks. After the little tree finishes blooming, transplant it into the garden where it can grow and rebloom for years, reaching a height of 6 to 7 feet with a spread of 5 to 6 feet in full sun. Either transplant the forced bulbs and hope they'll rebloom next year or add them to a compost pile. Reuse the container for another seasonal or permanent display.

opposite This painted wooden planter with footed base, miniature wire fencing, and spring blooms brightens a sunlit room for weeks.

HOW TO MAKE A TABLETOP GARDEN TO CURE CABIN FEVER

Use this miniature landscape as a centerpiece that brings hints of spring indoors with fragrant flowers and greenery. Set it away from bright light and heat so it will last as long as possible.

1 PROTECT Line the bottom of a wooden container with a sheet of plastic or set the planter on a tray to protect the tabletop from potential moisture damage.

2 LINE Cut cotton batting (available at a fabric store) to fit the inside bottom and sides of the container. Fit it into the box to hold moisture and soil. Water sparingly when soil feels dry. Do not let soil become soggy.

3 PLANT Add potting mix to the container then position the willow. Add more mix to about 2 inches below the rim of the container then snuggle bulbs into place.

4 FINISH Fill in between the bulbs with potting mix and then top with preserved moss.

Basket Gardens

Transform any basket into a mini garden. Line the interior, fill the basket with potting soil, and tuck in foliage or blooming plants.

Make room indoors for a basket of blooms or greens any time of the year. As a temporary home for plants, lightweight portable baskets work beautifully, blending well with many decorating styles. Baskets can be hung to highlight a vining or trailing plant. For planting, find a range of basket options—small or tall, round or shapely, spacious or compact, rustic or refined.

Baskets abound

You may already have a collection of baskets ready for planting. If not, economical options can be found at yard sales and thrift stores. Once you decide where you'd like to display a basket garden, choose an appropriate container for the setting. Place a large basket on the floor to accent an entryway, but keep the plants away from cold drafts. Place small baskets on tabletops, counters, or windowsills where they'll add beauty and interest.

Wicker baskets contrast with natural contents, whether green or flowering plants. Of course, not all baskets are woven with willow, jute, reed, wood, or other natural materials. You might find baskets made of metal, plastic, or concrete. Each material has its charms, depending on what appeals to you and suits your home. It's easy to change a basket to match or complement decor with a coat of paint or stain.

Waterproofing a basket

Most baskets do not provide a watertight container for plants. Protect the basket and extend its utility by using it as a decorative cachepot, lining it with sheet plastic or a saucer before planting. Most baskets deteriorate over time when exposed to moisture. Minimize water damage by brushing the basket with several coats of exterior-type polyurethane, allowing the sealer to dry completely each time before adding the next coat.

Planting a basket

Create a tabletop garden in minutes, combining common houseplants, annual flowers, or seasonal bloomers. Keep to a simple color scheme, group green plants with very different textures, or change out seasonal plants. Be bold and daring. Because basket gardens are temporary, you can easily change various elements of the combination for different looks.

Combine plants with similar light requirements. When using a basket as a cachepot for a combination of potted plants, plant selections might include ones with differing water needs. If you plant directly in a basket, first line it with sheet plastic, landscape fabric, sheet moss, or cotton batting to hold the soil, then water only enough to moisten the soil.

Design a basket garden by gathering potted plants and trying different groupings until you have a pleasing arrangement. Group odd numbers—three, five, or seven plants—for an effective display that can be viewed from all angles. When you're happy with the design, either leave the plants in their pots or remove the pots and sprinkle potting mix between the root balls. Variety and lushness are important, while leaving growing room between plants. For the easiest garden of all, fit a single plant, such as a lush fern, into a basket.

BHG TEST GARDEN TIP — QUICK GARDEN

Small baskets hold nursery pots of Siberian squill (*Scilla siberica* 'Alba'), daffodils, and small fritillaria in a spring tabletop garden that blooms for weeks. Sprinkle grass seed onto the soil between the potted bulbs to sprout a fresh green fringe within days.

BLOOMING BASKET

Fill a sturdy wire basket with enough fragrantly blooming plants to give visitors a colorful welcome that will leave a lasting impression. Set the basket on a saucer or tray to protect the surface beneath it or hang it at eye level for all to enjoy.

1 LINE
Use preserved sheet moss from a crafts store to line the basket and hold soil. Cover the interior—bottom and sides—of the basket, overlapping the moss.

2 PLANT
Remove plants from nursery pots and position them in the basket. This display features colorful primroses and aromatic hyacinths. Fill between plants with potting mix.

3 ENJOY
Park the spring-in-a-basket display next to a doorway where it can be most appreciated. Water plants lightly to sustain them during their flowery show.

USE A STRAWBERRY JAR CREATIVELY

Plant mail-order prechilled lily-of-the-valley pips (rootstocks) indoors then savor the prespring show of perfumed blooms in three to four weeks.

1 PREPARE
Soak the pips in warm water for 2 hours before planting. Meanwhile prepare the jar by adding first a layer of sand, then potting mix up to the first pocket.

2 PLANT
Carefully plant several pips in each pocket, pointed shoots up. Cover pip roots in each pocket with sphagnum moss. Cover the moss with potting mix.

3 DRAIN
Stand a section of drainage material in the center of the pot. Add potting mix to the top opening or next level of pockets.

4 FINISH
Plant any remaining pockets and the top opening of the pot. Cover the roots with sphagnum moss and potting mix. Water slowly and thoroughly.

5 MAINTAIN
Set the pot in a cool (60°F) location with indirect light from a north or east window. Water only to keep potting mix damp.

Creative Containers

Let your imagination run wild when it comes to houseplant containers.
Take cues from the plants themselves when choosing vessels.

Add to the fun of gardening indoors by finding and using unusual containers. The best containers are functional, providing space for roots to develop and plants to grow and flourish. Drainage holes or other means of releasing excess water are crucial to plant survival. Containers can be small enough to hold only one plant or large enough to accommodate a junglelike display of foliage and flowers.

Finding special containers

Indoor gardens benefit from a vast selection of potential containers and exotic plants. Even traditional houseplants, such as schleffleras, appear exotic when planted in effective containers. Make a style statement with any container garden, using planting schemes to express colors, shapes, and textures that fit your home. As you match plants with planters, consider how they will look in their setting. Colorful containers bring energy to green plantings, for instance; earth-tone planters meld with most settings. When choosing unusual containers for indoor gardens—whether long-term plantings or temporary projects—be wary of massing too many eclectic plant holders, which can result in a hodgepodge. Simple planters let plants dominate, yet improvisation encourages experimentation.

Treasure hunting

As seasons come and go, gardening interests and decorating styles go with them. Lacquered pots chosen for last winter's holiday plants may have lost your favor after a year or two in storage. New plants and different planting ideas call for fresh containers, and you'll find potential planters in many places, aside from garden centers and nurseries. Whether you're on the hunt for a vintage birdcage or a home for new geranium varieties acquired at a plant sale, look for container possibilities at thrift stores, rummage sales, auctions, and flea markets. Sometimes the funkiest junk—a rusty bedspring or bait bucket—makes the coolest planter.

Special containers

In addition to traditional pots and planters, you'll find less obvious but widely available ways to house plants. Self-watering pots, wall pots, and strawberry jars are only some special containers that work well for plants. Sometimes buckets, baskets, and planting pockets sold for outdoor gardening are easily adapted to indoor use and last longer.

Strawberry jars demonstrate the adaptability of classic garden containers. Glazed or unglazed strawberry jars have plantable pockets for growing multiple plants in a single pot. Planted indoors, lily-of-the-valley rootstock (rhizomes are called pips) can be brought into sweetly fragrant bloom weeks ahead of the gardening season.

BHG TEST GARDEN TIP

PICK A POCKET

Fill a strawberry jar with a medley of herbs, sedum, ivy, or scented geraniums. Young plants from packs or flats fit into the pockets easily, have room for rooting, and result in healthy plants.

opposite top The strong stem cuttings of dracaena, known as lucky bamboo, resemble a tropical grove rising from a sea of black polished stones in this copper wok. Over time, the stem cuttings will develop more leaves and a lush tropical appearance.

left and above Inspired by nature's rhythms, tabletop gardens that change with the seasons create interest and beauty.

Seasonal Centerpieces

Enliven seasonal tablescapes with greenery, natural elements, and a touch of whimsy.

Many gardeners take pleasure and find inspiration in anticipating the seasons, then plant as a celebration of each dynamic, distinctive time of year. In many households and families, seasonal celebrations are a focus. Dining tables offer platforms for changing displays of plants and accents, whether festive or casual.

Follow nature's lead

Find inspiration in the world around you, then arrange elements that appeal to you. When an orchid blooms, call on it as the centerpiece. When you plan a party, use small potted plants from the tablescape as take-home gifts.

To create seasonal displays other than tablescapes, consider a mantel, entryway table, or shelf. Following winter holidays, let ordinary cut flowers and evergreen boughs give way to potted spring-flowering plants and dwarf evergreen trees in a series of colors, textures, and scents.

Make it tidy

Keep the display fresh and clean. Although a view of exposed soil is tolerable, albeit not so appetizing, crumbs of soil won't be welcome in or near food. Tabletop arrangements should allow room for serving pieces and other dishes as well as unobstructed eye contact across the table.

January

Starting over Indoor bulbs and fresh blooms chase away winter dreariness, bringing cheer to short, dark days. Welcome the new year with ethereal paperwhite narcissus, bright pansies, and young dwarf myrtle (*Myrtus communis*). Candlelight adds a warm glow. *Photos opposite, top and bottom right.*

February

Romantic expressions Fill the indoors with the anticipation and excitement of the sweet-scented season. Mini roses and white cyclamen make lovely valentines; fresh-cut pussy willows warm hearts and encourage hopes for spring—and maybe a little romance. *Photos at right.*

Seasonal Centerpieces

MARCH

Spring returns Pots of posies, including primroses and crocuses, herald the arrival of the annual gardening season. Count on a local garden center, nursery, or florist to have seasonal selections of small economical potted plants that are perfect for tablescapes. *Photos at top left and bottom.*

JUNE

Casual gatherings Long days, strong light, and heat send gardeners into the cool shade. Find inspiration from under leafy green trees where you can sit back, enjoy the plants, and just breathe. Take time to relax and sip icy tea among verdant ferns and other green plants. *Photos below.*

AUGUST

Summer's end Make a place to unwind and eat fresh fruit at the end of a high-energy day. As summer gives way to fall, embrace simplicity during the season of plenty with a simple still life. Select a few garden treasures for sculptural qualities. *Photos at right.*

DECEMBER

Winter warmth Indoor gardens keep gardeners happy and welcome visitors with all the warmth and spirit of a cheerful home. Gather tiny evergreen trees, sparkling baubles, and plenty of candles to make merry. *Photos below.*

Topiary Towers

Precisely clipped topiaries are classic—indoors and out. Grab a pair of pruners to create stylish or whimsical shapes.

Experiment with the ancient art of shaping plants that elevates pruning to an art form. Topiary conjures a sense of elegance and tradition, yet requires little more attention than a typical houseplant. Indoors, a topiary provides a neat and portable focal point. Grouped topiaries of different plants, shapes, and sizes make a fascinating garden.

Plants for topiary

Start with a ready-made topiary and maintain its shape with periodic trimming, or sculpt a plant into a sphere, cone, spiral, animal shape, or other form. Either way, the easiest topiaries are those with simple geometric lines. Depending on your choice of plants, a preformed topiary frame can make shaping easy.

Moss-stuffed wire forms covered with small-leaf plants, such as ivy or creeping fig, are fun to use and to watch grow. Plants root into the moss and need to be kept damp and clipped regularly. To make two-dimensional topiaries, wind trailing or vining stems around a wire frame.

When selecting a plant for topiary, choose from herbs and evergreens with dense, compact foliage, including rosemary, lavender, santolina, scented geranium, myrtle, cypress, and boxwood that respond well to pruning. It can take a year or two to transform a healthy plant into a double-ball form.

Shaping topiary

To maintain a ready-made topiary, clip the tips of new growth, removing up to 2 inches to encourage branching. When training a new topiary, select a plant with a strong central leader or stem, then determine shape and ultimate size. Use floral scissors or pruners to trim plants. For a ball shape atop a main stem, start by pruning the lower branches to reveal the stem. The type of plant and the shape you want will determine how often you trim. Compact myrtle needs clipping every six to eight weeks; boxwood can be kept in shape with trimming three times a year.

Let topiaries rest in winter; lay off clipping and feeding for two or three months. Otherwise, fertilize a topiary monthly during the growing season. Take topiaries outdoors during summer, then move them back indoors to a sunny windowsill for winter.

ROSEMARY TOPIARY

Rosemary is an ideal plant to make into a standard or tree-form topiary. Its woody stem and needlelike leaves develop into a nifty little tree, and you can use the aromatic trimmings in cooking.

1 TRANSPLANT
Remove the plant from its nursery pot and give it fresh potting mix in a slightly larger pot of suitable scale.

2 STAKE
Cut a stake; push it into the soil next to the plant. Tie the stem loosely but securely to the stake using raffia or twine.

3 SNIP
Every few weeks during the growing season, snip stem ends to promote bushy side shoots.

opposite A pair of double-ball eugenia (Syzygium) topiaries enhance the setting with their sculptural forms. Similar ready-made topiaries of euonymous, lavender, or rosemary are widely available.

Support Systems

Some plants need a little help from their friends—stakes, poles, forms, and other structures—to keep them standing tall.

For vining and climbing plants that need support to look their best, make or buy pot-size trellises or tepees to hold them up. Vines may twine or cling to a support on their own with little encouragement. Other climbing plants require training to start them on a support and keep them growing. To grow a living curtain, use long bamboo poles, string, or monofilament to connect a windowsill planter to the top of a window frame and guide the stems.

Creative staking

Use plant supports creatively, rather than merely pushing a stake into soil and tying a plant to it. You might train a vine to an arching trellis or a topiary form, coax plants to climb into a living wall, or sculpt plants using an imaginative trellis. Sometimes staking provides a temporary solution, such as upholding a tall, large, or sprawling plant while it overcomes lopsided growth; later the stake can be removed.

When and how to stake

Stake or train plants when they are young and pliable. Prune the plant to cover its support as it grows. Training a plant onto a form takes extra maintenance. Be prepared to allow two years or more to develop an attractive topiary.

Choose a support that is at least as thick as the stem it will hold. Thin stakes break and flimsy tepees crumple, especially as the plant becomes heavier with growth. Stakes and other supports made of bamboo, bentwood, or metal—especially green or brown—blend in among foliage. If you want the support to add visible structure, make it white or a contrasting color.

If a plant cannot entwine or cling to a support on its own, tie it to the stake or trellis using a soft material, such as jute, raffia, or fabric ribbon. Wire and twist ties can injure plant stems. Tie loosely, giving the plant room to move and grow. As a plant grows, check the ties occasionally; remove and retie as needed.

Make a moss pole

Philodendron, Swiss cheese plant, and pothos are among the plants with aerial rootlets that prefer to grow on a damp medium such as a moss-filled post. Made by rolling a length of chicken wire into a 2- to 3-inch-diameter cylinder and filling it with sphagnum moss, the post can be anchored in a pot at planting time and become a permanent part of the plant/pot combination. Use hairpin-shape wires to train the plant stems to attach to the pole. Water the moss well at planting time and keep it damp with weekly misting.

BHG TEST GARDEN TIP

SUPPORTING ORCHIDS

Use a clip and a stake to keep slender orchid stems upright and sturdy. Stake a bloom spike when buds form, using a support about the same height as the flowering stem.

opposite, above right When needed by a plant for support, a trellis or stakes can hide among the foliage or add to the decorative nature of the display.

SCENTED TOPIARY RING

Scented geranium varieties with small leaves and a trailing habit, such as nutmeg and lavender, work well to make a lovely, fragrant topiary ring. Gather an 8-inch pot, two scented geraniums, and an 8- to 10-inch-diameter ring bent from heavy-duty wire or a clothes hanger.

1 PLANT
At planting time, set the plants at an angle to follow the ring—one right, the other left.

2 TIE
Use fine twine to tie the geraniums to the ring here and there. Add ties as the plants grow.

3 TUCK
Gently tuck in new growth or errant stems. Keep the plants in bright light.

CREATE A BONSAI

Potting a bonsai begins with preparing a young tree or shrub for its new home. Keep an indoor planting away from heat vents, fireplaces, and drafty windows and doors.

1 POT
Prepare the shallow pot for planting. Anchoring wires will help hold the plant in the pot.

2 COMB
To promote root growth, use a metal hook to comb out roots and loose soil.

3 CUT
Prune away excess roots, making the root ball compact enough to fit the shallow container.

4 WIRE
Keep the top-heavy tree from tipping out of the pot by wiring it in place.

Bonsai

The ancient art of bonsai offers beauty and balance in a pot. Try these zenlike plantings using a variety of houseplants.

Combining art with gardening, bonsai also mixes in a bit of magic to transform trees and shrubs into diminutive potted treasures. Bonsai—pronounced bone-sigh—originated in China more than 2,000 years ago, before being popularized in Japan. Although some bonsai specimens live for centuries and become heirlooms, it is possible to create a basic bonsai within a few years.

Beginning bonsai

Like many gardening endeavors, passion often propels an interest in bonsai to grow into a collector's hobby. The small plants are artistically trimmed and trained to appear ravaged by time, windswept, cascading, or draping a rock. Cultivating the miniaturized forms may seem intimidating at first, but some fundamentals will help you take up this garden art.

Plants: Choose a tree or shrub that suits your climate and lifestyle. Some plants need more maintenance than others. Cultivate trees such as evergreens, oaks, and elms outdoors. They need cold-winter dormancy to maintain their vigor, and must be protected from weather extremes. Some tropical and subtropical shrubs work well for bonsai and can be kept indoors year-round.

Start with an already formed bonsai, if you prefer, and practice annual root pruning in spring and regular branch tip pinching.

Pots: Bonsai live in shallow pots, usually at least three times as wide as the height. The container is part of a harmonious composition and it suits the plant's size and style. The pot must have at least one drainage hole.

Culture: Depending on the plant and the climate, some bonsai need intense light; others adapt to bright or moderate light. If a plant does not get enough light indoors or its location changes, it may drop its leaves.

An ideal potting mix for bonsai is porous, containing equal parts sifted bark and ground lava rock. Frequent watering—daily during hot weather—is necessary. Water when the soil feels slightly dry. Fertilize every other week throughout the growing season with diluted all-purpose plant food.

Repotting: Repot bonsai with fresh potting mix every two to four years, depending on the plant's needs.

Training and pruning: Guide the plant to grow into a graceful form by shaping the branches using copper wire and by pruning the roots, branches, and leaves. Shaping requires time and patience. You can create bonsai with one of many different styles, including formal upright, slanting, cascade, or forest (multiple plants).

BHG TEST GARDEN TIP — GOOD CANDIDATES

Some houseplants are suitable for growing bonsai indoors. Fig does well if kept in very bright light. Other options include schefflera, Japanese serissa, and dwarf pomegranate.

opposite, top An aralia branch—barely covered with soil—has sprouted and formed what appears to be miniature trees. For bonsai, choose plants with naturally small leaves, flowers, fruits, and branches.

Hanging Planters

Add vertical appeal to houseplants by suspending them with hangers or hanging them on a wall.

Take your indoor garden to the next level by raising plantings to eye height or higher. The plants will be easier to view and appreciate when separated. Hanging planters take advantage of vertical space that is usually overlooked. They can be suspended from a window frame, ceiling, or wall, which is an opportunity to create more beautiful displays of plants or enliven a stark corner. Also use hanging plants to highlight architectural details or to camouflage unattractive ones.

Ideal plants

The range of plants especially adaptable to hanging include those that trail or cascade (string of pearls, burro's-tail, grape ivy, creeping fig, ivy) and those that climb (jasmine, Madagascar jasmine, heartleaf philodendron). Naturally arching plants—holiday cactus, spider plant, and staghorn fern—also strut their best in a raised display. Choose a plant that will make a good impression as a focal point with its outstanding form, color, or fragrance.

Containers

Hanging planters also can present the challenge of water drainage. To prevent water from dripping on floors or furniture, choose containers that will catch any drips. Some pots have attached saucers. Water carefully then empty the saucer to avoid overflow. Alternatively, place a potted plant inside a plastic-lined basket or a decorative cachepot without a drainage hole, using lightweight containers suitable for hanging.

Hardware

Explore options for hanging plants from a window frame, ceiling, or wall, then take extra steps to secure suspended containers. Plenty of sturdy hooks, brackets, and other supports are available to handle the job. Install hangers and shelving by anchoring them into wall studs and ceiling joists, not into drywall or plaster. Choose hangers that have built-in swivels so you can easily rotate all sides of the planter toward the sun or light.

A container garden can become quite heavy with wet soil and plants. Only use supporting materials that can handle the weight. Fill planters with a lightweight potting mix instead of heavy potting soil.

BHG TEST GARDEN TIP

WATERING HANGING PLANTS

Consider ease of watering when you choose the height of the hanging container. Exposed to warm, dry air near the ceiling, hanging planters may need more frequent watering than floor-level plants. Find a watering system that allows you to leave the plant in place instead of hauling it to a sink.

opposite, left Fast-growing tradescantia forms curtains of living color. The 6-inch-diameter pots are not too heavy to hang from a sturdy curtain rod. ***opposite, top right*** Plastic-lined window box-style baskets hide lightweight nursery pots of the peace lilies. ***opposite, bottom right*** Small ceramic hanging pots hold rosemary and sage within easy reach in a kitchen window.

Forcing Bulbs

Coax spring bulbs into bloom and breathe in fresh scents and vibrant color during late winter.

You can fool Mother Nature, and flowering bulbs grown indoors prove it. Why wait until spring to enjoy the beautiful blooms of daffodils, tulips, and other bulbs when they can flower indoors weeks ahead of schedule? Forcing enables spring bulbs to bloom early during quiet months of winter. Force a variety of bulbs and savor an extended show with spectacular blooms, fragrance, and color.

Mark the calendar

Plan to force bulbs in a three-stage process that is easily managed over months. Purchase bulbs in fall and, instead of planting them in the garden as you would for spring blooms, plant the bulbs in pots of soil and chill them. Most types of hardy bulbs require chilling at 35°F to 48°F in a refrigerator or unheated garage or shed for at least 12 weeks.

When chilling bulbs in the refrigerator, store them away from fruit. Bulbs will not develop and flower when exposed to ethylene gas released by some ripening fruits. Label each bulb variety's completed chilling date, identifying required chill time, which varies from 12 to 14 weeks for hyacinths and squill; up to 15 weeks for daffodils, fritillaria, and grape hyacinth; and 14 to 20 weeks for tulips.

left Spring-flowering bulbs, including daffodils, hyacinths, and tulips, bring a blissful antidote to winter's shorter, darker days when forced into bloom indoors.

Planting and forcing

Fill 4- to 6-inch-deep pots, which have drainage holes and saucers, with all-purpose potting mix. Pot size depends on the type and quantity of bulbs planted. One large daffodil bulb or five crocus bulbs will fill a 4-inch pot; six hyacinth bulbs need an 8-inch pot. Plant the bulbs points up and poking through the surface of the potting mix. Avoid pressing on the bulbs and the soil to help the roots grow freely. Water thoroughly to moisten the mix. Keep plantings evenly damp, watering throughout the forcing process when the soil begins to feel dry. If potting mix is too wet, bulbs may rot or fail to grow well.

At the end of the chilling period, move the potted bulbs to a cool room and set them away from direct light. Keep them there for several weeks while roots develop. When bulb tips reach 1 to 2 inches tall, move the pots to a warmer room with more light. Within a few more weeks, the bulbs will begin flowering. Keep them out of direct sun to extend blooms.

After bloom

Forced bulbs cannot be forced into bloom again. But bulbs that have been forced indoors can be transplanted into the garden and may resume flowering after a year or two of recuperating. It's worth a try. After the forced blooms fade, cut the flower stalks (but keep the leaves), and continue to provide water and sunlight. In spring, transplant the bulbs into the garden, allowing the foliage to wither naturally, which feeds the bulbs' future blooms.

Easiest bulbs

To enjoy blooms more quickly, purchase pots of chilled, just-sprouting bulbs from a garden center or nursery for a delightful indoor garden. Just take home potted bulbs, tuck them into pretty containers (left in or removed from nursery pots), and watch them grow.

Varieties of paperwhite narcissus require neither soil nor prechilling for forcing. Using a container that holds water, such as a shallow bowl or wide vase, plant paperwhites in a potting medium of pebbles, glass marbles, or soil.

DON'T WAIT FOR SPRING

Paperwhite narcissus bloom within a month of planting. Plant them every few weeks for an ongoing display of intensely scented flowers.

1 PLANT
Place a 2- to 3-inch layer of pea gravel or other medium in a container. Set bulbs shoulder to shoulder.

2 COVER
Add water just to the bottom of the bulbs. Cover the bulbs up to their necks with pebbles.

3 ENJOY
For best blooms, keep the bulbs in a cool (60°F) room. Add only enough water to moisten the roots.

BHG
TEST
GARDEN
TIP

CHILLING REQUIREMENTS

Prior to chilling bulbs for forcing, identify the required chill time for each type of bulb.

Forcing Hyacinths

Easy-to-grow hyacinth bulbs offer a wide range of decorator options. And you'll love their strong perfume during winter months.

The marvelous, sweet scent of hyacinths triggers visions of spring. Like many bulbs, hyacinths can be grown in soil or gravel when they're being forced. But it's easiest to force hyacinth bulbs in water.

Traditional forcing

Forcing hyacinth bulbs has been popular since the Victorian era. People still use the traditional hourglass-shape forcing vases ideally suited to the task, with a narrow neck that holds the bulb just above the surface of the water. Various common household containers make excellent forcing vessels, as long as they hold water. A glass container—clear or tinted—is particularly desirable for displaying forced hyacinths because it shows the water level and the entire bulb as it develops. Hyacinth varieties bloom in hues of pink, purple, blue, apricot, yellow, and white.

Hyacinths—as well as crocuses and daffodils, such as tazetta and poetaz types—force well in the refrigerator. Chill hyacinth bulbs for 12 to 14 weeks. After chilling, you have the option of planting the bulbs in a pot with their tips poking out of the soil.

To force bulbs in water, set an individual bulb in a forcing glass. Alternatively, place a 3-inch layer of pebbles in a roomy vase or other watertight container then add water until it barely reaches the surface of the stones. Add a handful of horticultural charcoal chips to keep the water fresher longer. Set the hyacinth bulbs on the pebbles, with the bottoms at the surface of the water. Place the forcing glass or vase in a cool room away from direct sunlight.

After several weeks, the bulbs will show signs of growth as little green tips sprout. Add water to the container only as needed to reach the developing roots. Move the container into a warmer room with more light. Within a few more weeks, the bulbs will bloom. Gently transplant them into pretty arrangements, if you like. To use potted bulbs in arrangements, unpot them, gently tug their roots apart, and rinse the soil off the roots. The flowers last longest when displayed away from bright light and sources of heat.

opposite, top Bringing richly scented hyacinth bulbs into bloom in simple containers of water is a fun winter ritual. Plan ahead by purchasing bulbs in the fall. Find a wide variety of bulbs on the Internet from mail-order sources. *opposite, center* A vintage glass punch bowl makes a lovely centerpiece when brimming with hyacinths. The bulbs snuggle in a bed of clear recycled glass and white garden rock that mimic ice and snow. Move the centerpiece off the table during meals to prevent olfactory competition between fragrant hyacinths and aromatic food. *opposite, below far left* Place several inches of colorful marbles in the bottom of a glass pillar, add water to barely cover the marbles, then add hyacinth bulbs, and—voilà!—you've made a perfect home to grow hyacinths. *opposite, below right* This classic forcing vase holds a single hyacinth bulb. The bulb blooms about eight weeks after chilling begins. *right* Slender canning jars work as forcing glasses. Combined with a tepee of forced forsythia branches, top-heavy flowering hyacinth is less likely to topple.

Forcing Branches

Get a jump on spring by cutting branches from flowering shrubs and bringing them indoors to bloom.

Just when a colorful pick-me-up is most welcome in a winter-weary home, it's time to bring in branches of flowering shrubs and trees to coax into blossom ahead of the season. In the process of gathering branches for indoor arrangements, you'll get the additional benefit of accomplishing winter pruning. Forcing works well for many spring-bloomers, from apple to quince and serviceberry.

Forcing is easy

Late January through March is the usual window of opportunity to gather branches. Depending on where you live, begin snipping branches after plants have had their winter rest, or dormancy, for at least eight weeks of temperatures below 40°F. The later in winter in which branches are cut, the shorter the forcing time.

Select healthy, young branches with lots of flower buds (flower buds are typically larger and plumper than leaf buds). Choose branches from crowded sections of the plant where it will benefit from pruning, keeping in mind that you are removing some of the plant's spring display.

For a handsome display, start with cuttings that are 1 to 3 feet long. Snippets less than a foot long can also be forced and used to create beautiful small arrangements.

Depending on the variety of tree or shrub, buds will open in two to four weeks. Branches that can be forced to bloom in as little as two to three weeks include Eastern redbud, forsythia, honeysuckle, and pussy willow. Branches that need more time for forcing include magnolia, mock orange (*Philadelphus*), and rhododendron.

left Forsythia is one of the earliest-blooming shrubs and it forces easily. Cut branches as early as mid-January and bring them into bloom indoors within a few weeks.

Quick arrangements

Choose a sturdy vase for forced branches that include cherry, almond, and plum in a range of bright white to pastel blooms. Use a heavy vase for 2- to 3-foot-long branches—large displays can be top-heavy and prone to tipping over. Compose the essence of spring in a vase by combining flowering branches with pussy willow stems and cut tulips. Pussy willow branches need no forcing. Just cut them for indoor arrangements before the fuzzy buds are past their prime. Keep the branches out of water to arrest bud development.

> **BHG TEST GARDEN TIP**
>
> ## PUSSY WILLOW BRANCHES
>
> Common varieties of pussy willow, such as the small tree *Salix discolor* and large shrub *S. caprea*, are among the first signs of spring, and no forcing is required.
>
>

FORCING BRANCHES

Forcing spring-flowering branches means bringing well-budded prunings indoors and tricking them into flowering as if spring has sprung. Follow these steps, then wait for the buds to swell and show color. That's the time to move the branches into a warmer room and enjoy the blooms that will soon appear.

1 CUT
Use pruning shears to cut pencil-thick branches at least 12 inches long. Strip the lowest buds from the branches.

2 HAMMER
Use a hammer to crush cut ends of branches to prepare them to absorb water.

3 SOAK
Gather and wrap the branches in newspapers. Submerge the bundle in warm water for 12 hours to help the shoots wake from dormancy.

4 WAIT
Stand the branches in a bucket of water and set the container in a cool room. Change the water twice a week.

Succulent Wreaths

Wreaths made from succulent plants require little water and are a natural way to decorate indoor and outdoor spaces.

It's easy to mix succulents in an array of types and colors to craft an eye-pleasing tableau of textures and hues.

Plants for wreaths

Succulents are ideal for wreaths because they grow slowly. Top picks include sedum, aeonium, echeveria, kalanchoe, mother of pearl plant, and jade plant. To create a colorful and textural wreath, choose succulents of varying colors and leaf size.

Allow rooting time

After planting a succulent wreath, allow time for the cuttings to root into the moss. Place the wreath in bright light (not direct sun) for about three weeks. Gradually increase light levels to full exposure. In the hottest Growing Zones, protect plants from midday sun. Indoors, set your succulent wreath in a south-facing window, greenhouse, or garden room. Water the wreath during this time; don't allow the moss to dry out. After about eight weeks, the plants in the wreath will be fully rooted and the wreath can be hung.

Finishing touches

Hang your wreath using galvanized hardware to prevent rust. If you'll display the wreath on a wooden surface, such as a door, consider covering the back of the wreath with plastic attached with crafts pins to protect the wood from moisture.

Long-term care

Live succulent wreaths will change over time because the plants will grow. When you prune it to keep its shape, reuse the cuttings to start another wreath. Succulent wreaths can last up to five years with proper care.

opposite Textural and colorful, a succulent wreath can last for several years. If displaying indoors, hang the wreath in a south-facing window.

BUILD A WREATH

Mix succulents in an array of types and colors to craft an eye-pleasing tableau of textures and hues. Start with a ready-made sphagnum moss wreath or purchase a wire form and sphagnum moss liner separately.

1 TAKE CUTTINGS
Take succulent cuttings a day or two prior to creating your succulent wreath. Cut stems 1 to 2 inches long. Place cuttings on a tray overnight to let cut ends callus (form a thin layer of cells).

2 SOAK FORM
Submerge a mesh-covered moss wreath form in water. Allow the form to stand in water until fully saturated.

3 MAKE PLANTING HOLE
Poke each hole into the form using scissors.

4 INSERT CUTTINGS
Poke the succulent cutting into the wet form. Arrange the cuttings as close together as possible. Aim for density for the best appearance.

Tongiorgi Tomasi AN OAK SPRING FLORA

OAK SPRING GARDEN LIBRARY

MARTIN & CLINEFF

CHAPTER SIX

Gardens Under Glass

Inside a shelter of sparkling glass, plants get the boost they need to thrive indoors while providing a bird's-eye view of the action.

Moss Gardens

Soft and lush moss is available in a wide range of types and colors. Create a small garden in any see-through vessel with this textural plant.

Moss—and little more than that—makes a marvelously simple miniature garden with Zen appeal. Combined with a few stones, tiny pinecones, or well-placed handfuls of fine gravel, a moss garden is more minimalist than elaborate. Making a moss garden can be a contemplative process with a serene effect.

Mosses are primitive evergreen plants that form emerald carpets in the landscape. Hundreds of species can be found in North America, growing in creeping or mounding colonies. Use a single species of moss in a dish garden or combine mosses in a brocade of soft and nubby textures, depending on what's available. Lush green moss can add jewel tones to many types of container gardens. An all-moss garden is best displayed near eye level where it may be most appreciated.

Growing moss indoors

When growing moss in an indoor garden, start with potted plants from a nursery or collect and transplant moss from your yard. Gathering moss from areas such as parks, public lands, or protected areas is illegal. If you prefer, purchase fresh sheet moss from a florist or floral supplier then soak it in water for an hour before placing it in a container garden. Dyed and preserved mosses, available at crafts stores and garden centers, present more options that need no special treatment. Online sources offer live and preserved mosses.

Container-grown moss will thrive for several months indoors, given conditions that are moist, with acidic soil and high humidity, which are similar to those found in nature. After planting, press on the moss to establish good contact with the soil. Keep moss damp, not soggy, and place a live moss garden in indirect light.

opposite Composed inside a clear glass vase, a miniature garden includes tufts of green moss and gray lichen. The faded photo is secured to the outside of the glass.

UP ON A PEDESTAL

Convert a pedestal cake plate with a domed glass cover into a little moss garden. Enjoy the verdant scene any time of year, whether you use living or preserved moss.

1 PLANT
Place pieces of moss on the plate, mounding them over soil in places to mimic the contours of a natural woodland terrain.

2 EMBELLISH
Add a few rocks, twigs, pinecones, or pieces of bark to enhance the scene. Include a tiny violet or other plant, if desired.

3 SPRITZ
Spray the arrangement with water, then cover it with the glass dome. Mist the moss weekly or when it appears dry.

4 AIR
If condensation forms on the dome, lift off the cover for an hour or two until the humidity dissipates.

Tabletop Conservatories

A throwback to Victorian times, modern conservatories offer architectural splendor for flowering and foliage plants.

Miniature conservatories, also known as Wardian cases, have been favorites of cold-climate gardeners since Victorian times. The fanciful glass enclosures are downsized versions of the elegant greenhouses and conservatories often pictured in English gardens. Full-size glass and metal or wood structures remain fantasies for most gardeners, but miniature conservatories offer a reasonable alternative for smaller budgets. All it takes to make one of these gardens under glass is a conservatory and a few humidity-loving plants, such as ferns, dracaena, and nerve plant.

The Wardian case

These charming enclosures for indoor gardens were developed in the early 1800s by a London doctor, Nathaniel Ward, who wanted to watch a sphinx moth chrysalis develop in a closed jar. Ward was so enthralled with his project's success that he began growing plants of all sorts in glass enclosures. Before long, the plant hunters of that time who roamed the planet in search of exotic new plants, depended on Wardian cases to transport their finds with minimal watering and attention.

Today, miniature conservatories have staged a comeback with the resurgence of indoor gardening and the renewed popularity of terrariums. A stylish conservatory becomes a beautiful tabletop garden when filled with small seasonally blooming annuals or bulbs. It's also an ideal nursery for young plants, especially ferns and others that thrive in the humid environment created by the structure.

Many miniature conservatories are so architecturally captivating that they're used as a handsome backdrop for a few plants or other decorative objects. Some conservatories have built-in trays to plant in or set potted plants on.

Planting in a conservatory

If your conservatory doesn't have a built-in tray, set it on one of an appropriate size and cover the bottom of the tray with a 1-inch layer of pea gravel for drainage. Sprinkle horticultural charcoal over the gravel. Or, if the tray is at least 3 inches deep, finish filling it with potting mix for direct planting. Set the conservatory in diffuse light to prevent it from overheating and scorching the plants.

Maintain plants in a little conservatory by checking them periodically and watering as needed. Remove the top of the container to ventilate the plants every few days and to prevent fungal or mold problems in the humid conditions.

A long list of plants will thrive in the moist conditions of a conservatory, terrarium, or garden under glass. These plants need watering less often than they would in outdoor conditions or when grown ordinarily as houseplants. Favorite foliage plants include dwarf sweet flag (*Acorus*), baby's tears, papyrus (*Cyperus*), ferns, fig, nerve plant, and Scotch moss (*Sagina*); and flowering plants, such as begonia, impatiens, orchid, and primrose.

opposite, left Individually potted orchids thrive inside a Victorian-style Wardian case on a four-legged stand. Set on a pebble-filled tray in bright, indirect light, the flowering beauties hit their stride. ***opposite, top right*** A miniature conservatory gives plants a decorative showplace and a draft-free, moist environment. Add visual interest to the display by including colorful pebbles or flowering plants. ***opposite, center right*** Given room to grow on its own, a brake fern thrives under glass. When a plant outgrows its glass enclosure, it's time to swap it out, cut it back, or trim and divide it. ***opposite, bottom right*** A tall Wardian case with a built-in tray has room for several plants: maidenhair fern, arrowhead plant, artillery plant, and prayer plant. Vents in the top supply fresh air to the container.

109

Terrariums

Gardens under glass make beautiful tiny indoor gardens that are nearly self-sufficient.

A terrarium defies many of the typical challenges of gardening. Crafted in minutes, a miniature garden encased in glass rarely sprouts a weed. The need for supplemental watering is reduced by condensation forming and water droplets dribbling down to the soil, recycling moisture in the self-contained environment. The portable container allows you to move the garden into bright-but-indirect light in different rooms.

Magic under glass

A terrarium can be full of surprises, holding exquisite orchids, lichen-covered branches, or a collection of small compatible plants that like low light and high humidity. Children are drawn to terrariums as miniature landscapes where they can learn about gardening. All it takes is a few plants and an old aquarium or goldfish bowl. Easy-care terrariums can hold potted plants, making it a breeze to change the display whenever you want.

The container you choose gives the terrarium its appeal, from the sparkle of a giant clear-glass globe to the shapely charm of a lidded apothecary jar. A terrarium that's open at the top allows air to flow freely; an enclosed terrarium needs little watering, but regular opening to allow ventilation. Scour crafts stores, flea markets, and garage sales for containers, whether you prefer an oversize brandy snifter or an old pickle jar. Choose a vessel that you can fit your hand into for easier planting and maintenance. Always plant in a clean container.

left Nerve plant, lady slipper orchid, and fern cohabit in a reproduction apothecary jar. A butterfly made of feathers completes the scene and withstands the damp environment.

Planting a terrarium

Any terrarium might hold a few usually hard-to-grow carnivorous plants or tried-and-true dwarf and slow-growing houseplants that will appreciate low light and high humidity. Plants with bold-color foliage are easy to see inside the glass and contrast with other plants. Start with various-size plants for a grouping. Succulents, cacti, and other plants that like dry conditions rot rapidly in a closed or moist terrarium, but will do better in an open glass container in sterile sand layered with pebbles.

Make an ideal potting mix for a terrarium by blending two parts peat-based potting mix, one part sterile sand, and one part perlite or vermiculite. After planting, top the soil mix with moss, more gravel, or sand for a decorative effect.

Keep it growing

Lightly dampen the soil after planting, sprinkling water until it drips into the pebbles. Water an open terrarium when the soil appears dry—every 10 days or so. A closed terrarium may need watering every two weeks or more. Remove yellowed leaves and spent flowers as they appear. Periodically wipe the glass inside and out with a clean cloth.

MAKE A TERRARIUM

Prepare the best home possible for plants by covering the bottom of the terrarium with a layer of drainage material (glass beads, pebbles, or aquarium gravel) and sprinkle in a handful of horticultural charcoal chips.

1 DRAIN
Place glass beads about 1 to 2 inches deep in a clear glass container.

2 ADD SOIL
Layer potting mix 2 to 4 inches deep, depending on the size of container.

3 PLANT
Gently place each plant, tucking the root ball into the soil mix and covering it.

4 FINESSE
Use a makeshift long-handle tool to reach into a narrow-neck jar.

left Spacious containers, such as carafes, vases, and jars, will house long-lasting gardens. These showcase (from left) a mini guzmania bromeliad; maidenhair fern and club mosses; and club moss, button fern, and variegated false aralia. **above** Add color to your terrarium using recycled glass for drainage. Place a bottle garden in bright, indirect light and occasionally turn it slightly to help plants grow evenly. **opposite** A crystal decanter or vase would work as well as a small pitcher to hold a mini potted rose. Green moss disguises the pot and adds to the charm of this display.

Bottles and Jars

You don't need fancy glassware to create a terrarium. Any clear-glass container will do the trick.

Common containers, such as a cider jug, juice bottle, or other vessel with small openings or necks, are potentials for growing plants in small confines; however, they typically require more dexterity and customized tools than containers specially designed for terrarium gardens. A bottle garden can be one of the most successful terrariums possible, giving plants an ideal home in which they can thrive for years with little upkeep.

Bottle gardens became popular during the era of bell-bottom blue jeans and macramé plant hangers in the 1960s and '70s. Today, they join other forms of terrariums in the latest search for decorative indoor gardens that require minimal outlay and fuss. What better way to display a few plants than in a beautiful glass vessel that can make even ordinary plants stellar? As with terrariums, look for attractive glass containers of any size or shape. Choose clear over colored glass, which would filter some wavelengths of light needed by plants to be healthy. The smaller the opening and neck of a container, the trickier it will be to plant and maintain.

Planting a bottle garden

As with any terrarium, fill the bottom of the bottle or jar with sterile, inert drainage material: pebbles, sand, or glass. Roll a sheet of stiff paper into a cone shape and use it as a funnel to add those materials to the bottle. Add horticultural charcoal chips to absorb soil impurities. The rest of the planting process is different than planting usual terrariums.

Add a 3- to 5-inch layer of premoistened soilless potting mix, most of which are a combination of peat moss, fine sand, vermiculite or perlite, and added nutrients. As long as the mix is premoistened at planting time and kept damp, the bottle garden should do well. If you let the mix dry out completely, it can be difficult to remoisten.

Round up a few small moisture-loving plants that will fit inside the bottle or jar. Now fashion a planting tool from an 18-inch length of bamboo stake by attaching a small plastic teaspoon to the end of it. A length of stiff wire with a spiral or small loop twisted at one end can also be useful.

Carefully coax each plant through the opening of the container and use the planting tool to nestle the root balls into the soilless mix and cover the roots. Work slowly and patiently to arrange and settle the plants.

The right amount of moisture

Overwatering is the most common problem that occurs with bottle gardens and other terrariums because excess moisture does not drain out of the container. The soil should feel damp, not soaked. It's better for your terrarium to be a bit dry than too wet. Experiment until you get a feel for how much supplemental water is enough for your plants and how often it is needed. Of course, you may not be able to touch the soil inside a bottle or jar. Learning to distinguish the appearance of wet or dry soil mix is key to the well-being of the plants. Too little moisture, and plants will suffer or die; too much moisture fogs the bottle and plants can rot.

When the soil mix appears dry or there are no moisture droplets inside the container, add a tablespoon of water. If the soil still appears dry the next day, add another spoonful of water.

A child's nature-inspired terrarium teaches basic gardening lessons. Pint-size plants along with treasures from the backyard take up residence in a cookie jar-turned-terrarium.

Mini Biospheres

Create a mini eco system in a jar filled with tiny plants; it will dazzle children and adults alike.

Mini biospheres, complete with plants and added finds from nature, give intrepid gardeners the big rewards of a small world. Making a little biosphere is a quick and easy project. Tucking in a few treasures gathered from the yard or a nearby natural setting adds another layer of interest as well as an outdoor adventure. Prizes might include seed pods, bark, or lichen-covered twigs picked up from the ground. Avoid digging plants from the wild, gather seed pods only if several others are found nearby, and leave abandoned bird nests alone.

Making a terrarium

Even young hands can plant and manage a simple terrarium. Make the terrarium in any lidded glass or plastic container with an opening wide enough for easy access. Canning jars, lemonade jugs, and canisters are among easy-to-locate prospects. If the container has no lid, top it with a small clear-glass plate or plastic wrap.

When placed in the right location with bright indirect light, an encased ecosystem can survive on autopilot with an occasional peek. Condensation from moisture in the soil collects on the glass and dribbles down, keeping the potting medium moist in a continuous cycle. A bit of added moisture may be needed once in a while, especially if the lid is left off.

Terrarium troubleshooting

Follow these tips to keep terrarium plants in top condition:

If leaves turn yellow or leaf tips turn brown, allow the terrarium to dry out a bit by taking off the lid for a few hours each day for a week or two.

If plants are stretching to reach light and becoming spindly, place the terrarium closer to brighter light.

If leaves are touching the glass or falling off the plants, trim plants that have grown to reach the glass and remove spent leaves and flowers.

SIZE MATTERS

Start with small plants that won't quickly outgrow the terrarium. If you're unsure of potential size, check plant tags or consult an expert when selecting nursery plants.

TRAILING CLUBMOSS
(*Selaginella kraussiana*)
The cultivar 'Aurea' is a spreading tropical groundcover with chartreuse foliage. It needs high humidity.

GUPPY PLANT
(*Nematanthus gregarius*)
It's fitting for a fishbowl: a guppy with orange flowers and shining foliage. Pinch plant tips to prompt branching.

BEGONIA 'BETHLEHEM STAR'
Miniature rhizomatous begonias with colorful leaves thrive in the moist conditions of a terrarium.

EARTH STAR
(*Cryptanthus bivittatus*)
A striking bromeliad that makes an ideal companion for carnivorous plants, earth star requires little care.

STRAWBERRY BEGONIA
(*Saxifraga stolonifera*)
This beauty has silvery patterned foliage like a begonia and plantlets like a strawberry.

Glass Bubbles

Clear glass orbs and bowls offer contemporary good looks for easy-care air plants.

Make a chic, modern statement with a bubble terrarium designed to hang in a window or set on a table or shelf. You'll find various shapes for hanging, from oval to teardrop and round, as well as multiple sizes of clear glass globes to accommodate the plants you want to display. Use a bubble-shape bowl or spherical terrarium for tabletop displays.

Hang 'em high

Glass bubbles made for hanging have a hanger built into the tops. When suspended from the top of a window frame using a strand of invisible monofilament, a bubble-glass garden catches light and attention. Hang them at eye level where they can be viewed easily up close without getting bumped. A grouping of bubbles, whether the same or different sizes and shapes, has artful appeal and suits small spaces.

Tillandsias, sculptural bromeliads also known as air plants, live happily in glass bubbles and add exotic form to a creative display. The hanging type of a glass bubble also works well for young plants or treasured varieties that benefit from the extra warmth and moisture in this sheltered environment.

These small (3- to 8-inch) glass globes, teardrops, and other hanging vessels usually have one or more openings in the sides for planting. Handle plants carefully to avoid damaging them when planting. For an effective, carefree display, make a little bed of moss in a glass bubble and tuck in a couple of earthy treasures gleaned from your yard. Try adding an acorn, skeletonized leaf, sprig of berries—whatever works together and brings nature indoors.

Tabletop versions

Various styles of spherical containers work beautifully to make tabletop terrariums. They can also be displayed on a shelf and sustained with supplemental lighting. Goldfish bowls, bubble vases, and large spheres with side openings are among container options available from crafts stores, florist suppliers, and online sources.

Stretch your imagination when making a little garden in a glass bubble. Take advantage of the container's clean lines to plant a miniature floating garden using a single water plant such as water hyacinth. Or create a miniature landscape with a few tiny plants, a bit of moss, and a structural accent (a dollhouse-size chair, wheelbarrow, or animal figurine).

KEEP IT UP

Air plants displayed in glass holders prefer bright, filtered light—in an east or west window, or a few feet away from a south window. Take the plant out of the bubble for a weekly splash in a water-filled bowl, then return it to its place.

left Contemporary globe terrariums hold fuss-free tillandsias or air plants. These epiphytes need no soil, only bright light and occasional misting or a weekly plunging into water. ***top*** This 18-inch-diameter bubble, viewed from any angle, holds false aralia, dracaena, 'Sinbad' begonia, polka-dot plant, 'Ripple' peperomia, and rex begonia.
bottom A teardrop-shape bubble and a glass corsage bowl pair for a mini naturescape.

Cloches

Bell-shape glass cloches are classic and dramatic gardens under glass, creating focal points that also protect delicate plants.

Some plants deserve special treatment, such as placing on a pedestal. Using a glass dome or cloche and a pedestal to display plants takes them to a stellar level and also provides a nurturing environment. In addition to the aesthetic appeal of this type of glass-enclosed planting, a cloche works as a mini greenhouse for plants that prefer warmth and humidity when kept in a home where winter heating and summer air-conditioning systems can challenge the hardiest of plants.

Put a cloche to work

Glass cloches can serve in outdoor gardens. Also known as bell jars, cloches have been practical garden ornaments for centuries. Antique bell jars from Old English, French, or Early American gardens are rare treasures. But any contemporary reproduction brings the graceful character of traditional cloches to gardens today, sheltering seedlings from frost. Indoors, the glass dome can provide housing for cuttings and seedlings, complete with the extra warmth and moisture that prompts growth.

Prop up the cloche by placing a wooden block or similar object under the bottom edge to allow air in and heat and moisture out. Because cloches make a mostly airtight seal, if you forget to prop it open or remove it daily or weekly, depending on its contents, plants may perish.

opposite, far left A 16-inch-tall glass cloche sits on a bed of bark chips and sphagnum moss, housing a pretty (and hungry) group of insectivorous plants: 'Scarlet Belle' pitcher plant, Mexican butterwort, and Venus flytrap. Insects can enter the raised dome on their own or be fed to the plants. **opposite, top right** Kept temporarily under bell jars, young scented geranium standards—or tree forms in training—benefit from the moist setting placed away from direct light. **opposite, bottom right** This glass act features jewel orchid, arrowleaf fern, and 'Chocolate Stars' earth star on a Victorian-style cake stand. Moss stabilizes and hides the soil and plant roots. **right** Seedlings thrive under glass cloches when placed away from bright light to avoid baking the tender young plants. Raise the jar on blocks or wood pieces to allow airflow.

CHAPTER SEVEN

Indoor Edibles

A surprising array of delicious and nutritious plants can make themselves at home indoors. Simple methods make it happen.

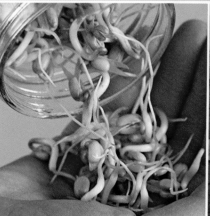

Herbs

Clip flavorful leaves from indoor herb gardens to use in sauces, baked goods, and teas.

When grown in containers, herbs bring indoors all the benefits of garden plantings—beauty, fragrance, flavor, well-being—plus convenience. Among the traditional outdoor plants that fare well indoors, some herbs excel in a sunny window, yielding incomparable fresh flavors for cooking while serving as delightfully aromatic houseplants. Herbs and other edible plants provide surprising beauty as well as bounty indoors, and it's simply fun to see how many ways you can use and enjoy them.

Growing herbs indoors

The culinary herbs used in cooking are ideal for a kitchen or dining room with a super-sunny window that faces south or west and provides direct, bright light. Most herbs need at least 6 hours of sunlight daily; supplemental light can help.

For good results, start with young plants. Grow them in a window box or individual pots on a sill or group them in a hanging basket. Annual herbs, such as basil, cilantro, and dill, will be challenged to survive a long, cold winter where the light is not always bright, but they may last for a couple months. Plants started from seeds will grow nicely indoors throughout the winter. To bring herbs indoors from your garden, the best candidates are tender perennials, such as rosemary, bay, and lemon verbena—especially plants that have spent the summer growing in containers.

To help keep herbs growing indoors, pinch the stem tips often. Plant growth slows naturally during winter, but progresses and quickens in spring. Popular uses of herbs in the kitchen include snipping fresh leaves into a salad, omelette, or spreadable cheese. Their aromas and flavors enhance any tea or fruit drink. Harvest no more than one third of a stem's length to keep the plant growing productively.

GROWING HERBS ON THE SILL

Create a snippable indoor herb garden in a sunny location in your home. Culinary herbs provide extra-special pleasures with flavors and fragrances.

1 GATHER
Start with young plants, terra-cotta pots and saucers, and potting mix. Transplanting takes less than an hour.

2 TRANSPLANT
Remove plants from nursery pots and slip them into the terra-cotta pots, adding fresh potting mix.

3 FINISH
Water well after transplanting. Avoid overwatering herbs by allowing the soil to begin drying between waterings.

Herbs around the house

Think of other ways to use herbs around the house. Group small potted plants on a dining table for an aromatic centerpiece. Let shapely herb topiaries accent a tabletop with their elegant charms. Flowering lavender plants serve as delightful, albeit temporary, pleasures indoors. Deck your house with large pots of rosemary and bay laurel as cheerful winter decor. Embellish the plants with sprigs of evergreens and bright baubles. To get the most benefit and pleasure from growing herbs indoors, make them part of everyday life, rather than reserved only for special occasions. Try gently squeezing a few basil leaves to release their fragrance momentarily or tucking a sprig of rosemary into a hand-penned note. Simple pleasures can be the sweetest.

opposite Some culinary herbs grow especially well indoors on a sunny windowsill. Good options include basil, oregano, parsley, and thyme.

Plants from Produce

Grow indoor edible gardens from seeds, pits, and roots of foods.

You may find treasure—and some gardening fun and adventure—in your kitchen trash or compost bucket. The beginnings of plants are contained in the seeds, pits, roots, and other seemingly useless parts of edibles. Instead of tossing them, consider growing the plants (shown opposite), as well as many others that grow readily from seed, tuber, or rhizome. Date, ginger, fig, strawberry guava, loquat, and others with less-familiar names have potential for growing in a windowsill garden. Citrus of all varieties sprout easily from seed.

Almost any unprocessed vegetable or fruit has potential as a fascinating houseplant, whether it grows as an annual, perennial, shrub, tree, or vine. The goal isn't an orchard, but a few houseplants to provide greenery and maybe some produce.

Plants once considered exotic—papaya, mango, kiwi—are now available year-round in supermarkets and furnish some of the most satisfying gardening adventures. Fair game includes any exotic fruit that you bring into your kitchen that has not been cooked, pickled, or irradiated (rendered sterile and won't grow). But you don't need tropical conditions or great skill to succeed. Some of the nation's most experienced produce-sprouters belong to the Rare Pit and Plant Council (aka "the Pits") and hail from the Big Apple. (Where else?) They meet in members' high-rise apartments where their garden experiments grow. Sometimes the plants require patience while waiting for them to sprout—some seeds take up to 12 weeks to germinate—but anyone can enjoy the process and keep the plants going from year to year.

left Even inexperienced gardeners can pluck a seed from fruit, poke it into soil, and enjoy the seed-starting magic that ensues. Next time you go grocery shopping, consider sweet potato, lime, passion fruit, or papaya for garden potential.

Growing exotic plants

Start plants from roots, bulbs, tubers, rhizomes, cuttings, or seeds, as shown here. Ginger grows from a knobby rhizome. A pineapple plant starts with a fresh-looking leaf rosette from the top of a ripe pineapple. Many seeds sprout from ripe fruit within one to eight weeks. Remove seeds or pit from the fruit and rinse well to remove any flesh. Soak the large seed of a mango in water for two or three days, then pry open the softened outer husk and remove the inner seed for planting.

Start seeds in damp potting mix or soilless mix. Cover the container and set it on a heating pad to help prompt germination. As seedlings begin to develop leaves, uncover them. Place the seedlings under a grow-light or in a warm sunny window and keep the potting mix damp. When seedling roots outgrow the original container, transplant them into a slightly larger pot using compost-enriched potting mix.

Keep plants in bright light, such as a south-facing window or sunroom, or use supplemental lighting to keep them growing strong. The plants will need a warm (65°F–90°F) location. Move tropical plants outdoors for the summer, where they will thrive in the sun and fresh air—rainfall and humidity are bonuses.

BRING THE TROPICS HOME

Start a collection of homegrown tropical plants and raise these exotic wonders in containers on a warm, sun-drenched windowsill.

1. AVOCADO Peel the papery skin of an avocado (*Persea americana*) before planting. A tree grown from a pit rarely fruits; purchase a grafted clone to produce fruit.

2. GINGER *Zingiber officinale* reaches 3 to 4 feet tall and will bloom in summer, if you're lucky. The plant drops its leaves and rests from late fall through winter.

3. PAPAYA The maplelike leaves of *Carica papaya* make it a handsome tree. The plant grows quickly but most likely will not bear fruit indoors.

4. PASSION FRUIT *Passiflora edulis* vines are self-fertile and reach fruiting size one year from sprouting seeds. Passion fruit produces exquisite blossoms in bright light.

5. PINEAPPLE Twist the leafy crown off the fruit and pull off the bottom leaves before planting *Ananus comosus*. Watch for fruit after two years.

6. POMEGRANATE Slow-growing *Punica granatum* may not fruit indoors but it makes a lovely plant. Pinch the stem tips of a young plant to prompt branching.

7. STARFRUIT The feathery leaflets of *Averrhoa carambola* fold at night. The tree grows about a foot a year.

8. SWEET POTATO Long a favorite of indoor gardeners, the tubers of *Ipomoea batatas* sprout in water or soil and produce quick-growing vines.

Citrus

From kumquats to limes, you can grow citrus for drinks and recipes. As a bonus, citrus flowers emit sweet fragrance.

Yes, you can grow citrus trees indoors, and dwarf varieties are particularly well-suited to pot culture. Worthy candidates for homegrown citrus include calamondin, kumquat, Clementine tangerine, Mexican (key) lime, 'Meyer' lemon, and myrtleleaf orange. Citrus plants offer something for all seasons: glossy dark green foliage, sweet-scented blossoms, and colorful, long-lasting fruits (depending on maturity of the fruit). The flowers will fill a room with fragrance. Many citrus plants are available in striking variegated forms, which bloom and fruit as readily as plants with all green leaves.

Where to begin

Start with dwarf-variety plants that will grow 3 to 6 feet tall indoors. Varieties other than these would grow to 25 feet outdoors, reach up to 10 feet indoors, and require regular pruning. Set plants outside to summer on a deck or patio or in the garden.

It's possible to grow most citrus from seeds salvaged from fresh, ripe, store-bought fruit. But most supermarket fruits are produced by hybrid trees, and plants grown from their seeds will not likely fruit. Or the tree may flower after many years, but the fruit may not be the same as the seed source. Nonetheless, citrus trees are lovely houseplants, whether they produce edible fruit or not. Sometimes the fruit of indoor-grown trees is sour or bitter when eaten fresh, but it can be used to make luscious marmalade or candied fruit.

left Potted citrus trees bring a touch of the tropics to a sunny room, especially during winter. Among the best choices for container growing indoors are the calamondin, dwarf orange, and 'Meyer' lemon.

What citrus plants need

Adequate light is essential when growing citrus indoors. The trees will grow well in a sunroom or conservatory. Otherwise, situate them near a bright south-facing window where temperatures range between 60°F and 85°F. Gradually acclimate a tree to more or less light when moving it outdoors in late spring and back indoors in early fall. Once outdoors, progressively move a tree into more and more light for about three weeks. Reverse the process in fall to help a tree adjust to less light indoors. Citrus are sensitive to cold and should be protected from frost. Also protect trees from harsh outdoor wind.

Pot a tree in a high-quality potting mix enriched with pine bark, sphagnum moss, and gradual-release plant food. Water consistently to keep the soil damp. Fertilize throughout the growing season. Give the plant a regular shower to clean dusty leaves. Prune citrus after flowering to shape the tree and to remove any dead or damaged wood.

YOU BE THE BEE

Citrus trees rely on bees for pollination. Ensure fruiting indoors by aiding the process. Use a fine paintbrush to transfer pollen grains from the stamens of a flower to the pistils in other flowers.

REPOTTING A LARGE PLANT

Citrus trees grow well in containers when their roots are constricted. If the plant wilts between waterings and needs watering more than once a week, it's time to repot. Otherwise, repot annually.

1 GET SET
Start with a pot 15 inches wide and deep for a young tree. When repotting, increase the pot size by no more than 25 percent.

2 UNPOT
Gently tip over the plant and slide off the old pot. Handle the tree by its root ball or, if you must, grasp it at the base of the trunk.

3 RELEASE
Loosen the root ball with your fingertips if the roots have become packed or tightly wound. Dislodge old loose potting mix.

4 REFRESH
Fill the bottom of the new pot with fresh potting mix. Set the tree in the pot and fill in around the root ball with the mix. Water thoroughly.

GROW YOUR OWN SPROUTS

Growing fresh sprouts at home requires no fancy equipment. Take a couple minutes to tend the seeds each day, and you'll be munching nutritious, flavorful sprouts in four to six days.

1 COVER
Place 1 to 4 tablespoons of prepared seeds in a clean, sterilized wide-mouth jar. Top the jar with fiberglass mesh and screw on the metal canning jar band.

2 RINSE
Use lukewarm water to rinse the seeds. Swirl and drain.

3 SOAK
Add about 1 cup of lukewarm water. Soak the seeds for 12 to 24 hours.

4 DRAIN
Pour off the water. Rinse and drain seeds twice a day, pouring off all excess water.

5 GROW
Keep the jar at room temperature. Rinse and drain twice daily as sprouts develop.

Sprouts

Enjoy this health food that's delicious, easy, and fast to grow at home. Raise your own sprouts from a variety of different types of seeds.

Germinated seeds or sprouts are one of the quickest, easiest, and most delicious crops for indoor gardeners. The most commonly grown sprouts, including alfalfa, radish, lentil, azuki bean, mung bean, red clover, beet, buckwheat, and broccoli, are widely available from supermarkets, food co-ops, and health food stores. But they often come with a premium price, and, by comparison, it's so inexpensive to grow them on your kitchen counter. There is neither soil nor windowsill involved. You can start with a packet of organic seeds or buy bulk seeds from an online source for the most economical approach.

Even if you haven't eaten sprouts, perhaps you recognize them as a healthful food. Sprouted seeds pack a lot of nutrition, are typically high in protein, vitamins, and minerals, and are low in fat and calories. Sprouts add tasty fresh crunch to salads and sandwiches. They can also be cooked in soups, stir-fries, breads, muffins, and more.

How to grow sprouts

The most common method of sprouting seeds requires a pint or quart jar, moisture, and plenty of air circulation. It takes less than a week to grow most sprouts. You can easily keep successive crops going. Taste sprouted seeds as they develop and use them when you best like the flavor. When sprouts are 1 to 2 inches long, depending on the variety, rinse them thoroughly before eating. Drain unused sprouts and refrigerate them in a covered container up to one week.

Avoid salmonella poisoning from raw sprouts by starting with certified pathogen-free seeds and following these guidelines:

Treat seeds before sprouting. Heat to 140°F a solution of 97 parts water and 3 parts hydrogen peroxide. Immerse the seeds in the heated solution for five minutes.

Place seeds in a container and add water to cover them plus 1 inch. Skim off floating seeds and other debris.

Sprout seeds using a clean jar and a new screw-top canning band washed with soap, rinsed, and sterilized with boiling water. After sprouting, wash the jar and lid, then disinfect them using 3 tablespoons bleach per gallon of water. Use new, clean mesh for each batch of sprouts. Or use jars specifically designed to grow sprouts.

opposite, top Just-sprouted alfalfa seeds, with barely developed leaves, have loads of nutrients. To green up sprouts, leave them without a cover for a few hours in bright but indirect sunlight. *right* Sprinkle bean sprouts onto any dish to add color and nutrition.

Microgreens

Tasty, nutritious, and easy to grow, this tiny crop is available in just a few weeks after planting.

Microgreens are very young seedlings with their first pair of true leaves. Grown for harvest at this teeny stage, microgreens are tasty and nutritious enough to elevate salads, appetizers, and other dishes to gourmet fare with a snip and a sprinkle. Chefs have bolstered the itty-bitty greens to the hippest status among food trends. Home gardeners have caught on, as the freshest of produce is one of the easiest windowsill crops.

Microgreens are not sprouts (germinated seeds, eaten root and all). If left to grow, microgreens become seedlings then full-fledged plants. Many vegetable and herb seeds produce delicious young leaves and stems that are ready for harvesting as microgreens by snipping the stems after two or three weeks. Unlike sprouts, these plants need soil and bright light to grow. Microgreens have more developed flavors than sprouts, and they're often grown for color, variety, leaf shape, and texture.

Superb variety

If you sow a different type or two of microgreens every other week, you can enjoy an ongoing supply of microgreens. Yet even after a year, you would have sampled only a fraction of the realm of possibilities. Among lettuces and salad greens alone, there are dozens of mild and tangy options, from arugula to spinach. Each variety of microgreen has a distinctive flavor and appearance—just as the full-size plants do. Choose your crops of microfennel, microcabbage, or anything else with thoughts of how you might use them: as gorgeous garnishes or distinct flavor ingredients. Consider sowing seeds of purple basil, red amaranth, and 'Bull's Blood' beet just for color. Or customize a blend of microgreens, sowing a mix of seeds with similar rates of germination.

Growing microgreens

Sow seeds in a container with a 2- to 3-inch-deep layer of premoistened soilless seed-starting mix or potting mix. Sprinkle the seeds evenly, leaving up to $\frac{1}{4}$ inch of space between them. Even microsize plants need room to grow. Cover the seeds with $\frac{1}{4}$ inch of vermiculite. Water gently. Cover the container with clear plastic until the seeds germinate. Remove the cover and set the container in bright light. A south-facing window is best; supplement with full-spectrum lights if need be. Keep the planting medium damp by watering from the bottom.

Most seeds will germinate within 4 to 14 days, and will be ready for harvest in another week or two. Winglike cotyledons form first, followed by a pair of true leaves. Use the microgreens soon after harvesting. They'll keep up to three days, refrigerated in a plastic food storage bag. If allowed to grow on, the microgreens will develop into seedlings of transplantable size. Some cut-and-come-again lettuce varieties will regrow; otherwise the crop is finished once cut. Reuse the seed-starting medium several more times then replace it.

opposite, left Many types of vegetable and herb seeds produce crops of colorful microgreens, perfect for winter salads. The seedlings, such as those of beet, radish, basil, and kale, have mild flavors compared to their full-size counterparts. *opposite, top right* Use sharp scissors to snip the stems a bit above soil level. Cut as much as you want; harvest the rest later. *opposite, center right* Mix several types of microgreens for a colorful, flavorful treat. *opposite, bottom right* A simple appetizer, such as a stuffed mushroom, becomes gourmet fare with a sprinkling of microgreens and edible flower petals.

131

Wheatgrass

Whether you juice it for your favorite smoothie or grow little pots for tabletop decor, wheatgrass is a fun crop to grow.

Indoor gardeners know that anything green and growing provides a sure cure for cabin fever. Sowing a tray or small pots of fresh wheatgrass brings an early dose of spring within days. Wheatgrass has several popular uses. It forms a sophisticated, but minimal, tabletop garden. Cut it and process it in a juicer to make a health drink. Cats munch on wheatgrass during winter for an indoor lawn of sorts; felines will lie on wheatgrass as well as nibble it.

How to grow wheatgrass

To plant a 10×20-inch flat of wheatgrass—(enough to make fresh juice every day for a couple weeks or a generous-size centerpiece). Or to plant in small individual pots or other containers, follow these instructions.

Prepare the wheatgrass seeds: Place 1½ cups of hard red winter wheat (*Triticum aestivum*) in a large jar. Fill the jar with water and soak the seeds for 12 hours or overnight. Use a fine-mesh colander to strain the seeds, then rinse them thoroughly under running water.

Plant: Prepare a seed-starting flat by adding a 2-inch layer of potting mix. Completely moisten the mix with warm water. Sprinkle the presoaked seeds over the potting mix, distributing them as evenly as possible. Sprinkle with warm water to help seeds and potting mix make good contact. Allow excess water to drain. Set the flat on a tray or in another flat that does not have drainage holes. Cover the seeds with a single layer of damp paper towels.

Set the flat in a warm place with indirect light for a day or two. Mist the paper towels with water daily to keep them damp. When the seeds germinate and sprout, remove the paper towels. Continue misting with water daily to keep the potting mix damp.

Harvest wheatgrass in 7 to 10 days, when it has reached 4 to 6 inches tall. Use sharp scissors to cut the grass. Harvest only what you will use fresh to make juice, cutting close to the soil. Or trim grass evenly to one height to use it decoratively. Either way, the grass will grow again before eventually deteriorating.

DECORATING WITH WHEATGRASS

Large trays and small pots of wheatgrass have decorative appeal. Wheatgrass grows easily and quickly enough to make a large quantity of tabletop gardens for a wedding or other special event. Small pots work well as name-card holders. Line a mantel or windowsill with small pots. Plant Easter baskets with wheatgrass (line the basket with landscape fabric before adding soil). Use a tray or a large low-profile container of wheatgrass as a centerpiece. Tuck in glass votive holders, cut flowers, or other embellishments to suit a decorating theme. A wheatgrass centerpiece never fails as a conversation starter. After 7 to 10 days, snip the grass evenly to keep it lush; cut it a couple more times over several weeks before replacing it with new wheatgrass.

opposite It's easy to maintain a steady supply of fresh wheatgrass. It germinates quickly and can be used to make nutritious juice.

CHAPTER EIGHT

Decorator Favorites

For gardeners of all skill levels, these beauties offer plenty of options for celebrating every season with color, fragrance, or texture.

African Violet

This furry-leaf favorite sports a wide range of beautiful blooms and adds instant charm to any windowsill or tabletop.

Once thought of as a favorite houseplant of grandmothers, African violet (*Saintpaulia ionantha*) offers an ideal experience for any indoor gardener. Few plants match African violet's ability to thrive and bloom indoors for months on end. Rosettes of fuzzy foliage and velvety leaves give violets a cuddly quality that has attracted admirers since the plants were discovered in eastern Africa in the late 1800s and introduced commercially. The plants, not true violets, are instead members of the *Gesneriacae* family and cousins of gloxinias.

Good growing

Diversity has sparked the popularity of African violets as collectible flowering houseplants. But violets are more than a seasonal pick-me-up. With proper care, the plants can bloom continuously and live for decades. Plenty of bright, indirect light is essential for an abundance of blooms. If natural light is too strong, the foliage can burn and grow extremely compact and brittle. If light is too weak, new growth will be spindly and no flowers will appear. Violets need a warm room (65°F–80°F) and ample humidity. Many African violet enthusiasts attest that 12 hours of fluorescent light daily is key to vivid color and large blooms.

Overwatering commonly spells the end of African violets. Water from the top or bottom, but only when the soil feels dry. Use room temperature water and avoid dripping water on the leaves, which can cause spots. When using a saucer, self-watering pot, or other bottom-watering method, do not let the plant sit in water for more than 20 minutes.

left Modern African violets impress with showy characteristics. 'Kaylih Marie' boasts variegated leaves. 'Ultra Violet Saturn' has white-and-magenta flowers. *opposite, center left* Diminutive African violets can be served up in a teacup.

Potting and other pointers

Hold off on repotting a new plant after you bring it home. Let the plant adjust to new surroundings. Repot only once or twice a year. African violet likes to be potbound, which allows the plant to put its energy into blooming. Choose a pot that's one third the diameter of the plant, whether it's a standard variety that can reach 10 to 12 inches in diameter or a miniature that grows less than 6 inches in diameter. Plastic pots work well because they help keep soil evenly moist. Use a lightweight potting mix of equal parts sphagnum moss, vermiculite, and perlite. A plastic pot lends itself to pretty display options because it can be set inside a decorative container.

A diluted dose of water-soluble fertilizer (a 15-30-15 formula at the rate of ¼ teaspoon per gallon) when watering will boost plant health and keep it blooming. Remove spent flowers and leaves. Remove dust or debris from each leaf using a soft-bristle brush.

CHEERS FOR *CHIRITA*

As an alternative to African violet, try growing its Asian cousin. Chirita (*Chirita sinensis*) is distinguished by its trumpetlike blooms and its needs are similar to those of African violet.

SECRETS TO SUCCESS

African violets thrive in a stable environment, especially given a warm setting with morning sun and no chilly drafts. Try these tips for making plants happy, producing more blooms, and starting new plants.

WATER
A long-spout watering can and a pebble tray make it easy to give African violets the water and humidity they prefer.

REPOT
Use fresh, well-balanced potting mix when repotting plants. Make the mix yourself or use a commercial mix formulated for African violets.

PROPAGATE
It's a snap to start new plants from leaf cuttings. Start with a healthy leaf cut at the base of a stem. Keep cuttings warm and damp.

Iron cross begonia has distinctive puckering and a dark cross-shape pattern on its apple-green leaves. It prefers to grow indoors and can be finicky about care.

Begonia

You'll love the puckered multicolor leaves and petal-packed lavish blooms of this easy-care houseplant.

Diverse begonias bring tropical flair to any setting, whether grown for distinctive flowers, showy foliage, or both. Growth habits range from tall and upright to bushy, creeping, climbing, tuberous, or trailing. Most of the plants can grow and bloom year-round in a conducive climate, although others must have a period of rest or dormancy during winter; some bloom intermittently. None tolerates cold.

Meet the begonias

Explore this plant family by starting with a type of begonia that appeals to you. Begonias are differentiated into three main groups. Rhizomatous begonias have creeping stems that root as they grow, and include rex begonia (*B. rex-cultorum hybrids*) and iron cross begonia (*B. masoniana*). Another group, which grows from a fleshy, bulblike stem base or tuber, includes tuberous and semituberous begonias. The fibrous-rooted group includes the hardiest cane-stemmed begonias (*B. selections*), shrublike begonias (*B. selections*), and wax begonias (*B. × semperflorens-cultorum*).

Growing begonias indoors

It's easy to find a begonia for a warm (65°F–70°F) room with bright, indirect light—usually in an east or west window. Varieties with red leaves do better in higher light levels. Plants stretching toward a window need more light; plants with scorched leaves should be moved into less direct light. Begonias in terrariums can adapt to a north window; others fare well under fluorescent lighting 14 hours a day.

Repot a begonia in peat-based potting mix when its roots fill the pot. Let the surface of the soil mix dry between waterings. Set the pot on a pebble-filled saucer to help prevent roots from sitting in water and rotting. Water plants less often during winter. From spring through fall, feed plants weekly when watering, using a quarter-strength fertilizer.

Start new plants from pieces of rhizome, leaf, or stem cuttings. Pinch stem tips during the growing season to promote bushier growth, but avoid removing flower buds. Allow potted tubers to rest over winter, storing them indoors in a cool, dark place and watering only occasionally.

BEGONIACEAE FAMILY

As representatives of the large begonia family, which includes more than 1,000 species, these are among the most common:

TUBEROUS Start the tubers indoors, but grow the plants of *B. × tuberhybrida* outdoors and enjoy spectacular flowers such as 'Nonstop Mocha Yellow' (shown).

ANGEL WING The leaf shape gives angel wing begonia (*B. coccinea*) its name. With dangling flower clusters, varieties such as 'Lana' (shown) can grow to 4 feet and become shrubby.

REX Fancy-leaf begonias (*B. rex cultorum hybrids*) such as 'Houston Fiesta' (shown) are grown for colorful patterned markings. They like high humidity.

RIEGER A type of elatior begonia, rieger begonia (*B. × hiemalis*) grows low and bushy with green or bronze leaves and bright color camellialike flowers.

CANE-STEMMED Tough, easy-to-grow plants include 'Dragon Wing' (shown). It blooms intermittently throughout the year, given enough bright light.

Bromeliad

For an exotic addition to your interior, look no further than spiky, colorful bromeliad.

Among the most architectural of houseplants, bromeliads make dramatic statements with exotic, sculptural forms and spiky textures. Some have colorful bracts or specialized leaves that resemble flowers. Bromeliads bloom, usually when the plant is 3 to 5 years old, then decline slowly. Before it dies, the plant produces offsets or pups—the next generation of plants. When buying a plant, choose one that has yet to bloom.

Unusual lifestyles

Bromeliads' adaptability, resilience, and comparatively easy care have bolstered their popularity in homes and offices. You can move the plants outdoors over the summer. They'll appreciate the fresh air and humidity, but need some shelter from sun.

Bromeliads prefer small, shallow pots. Because they can become top-heavy and topple easily, set a small pot inside a larger one. A bromeliad planted in a too-large pot can easily rot.

In the rainforests of South America, high deserts of Mexico, and other native lands, bromeliads grow on trees, rocks, other plants, and in the ground. The plants are not parasitic, regardless where they grow. These plants are grouped as follows:

Terrestrial bromeliads have roots, which enable them to grow in soil. This group includes the best-known bromeliad, pineapple (*Ananas*), as well as earth star (*Cryptanthus*) and dyckia. These plants are not efficient at absorbing water and nutrients through leaves. Plant them in an appropriate potting mix and water as needed. Light needs vary.

Tank bromeliads are mostly epiphytic, living on tree branches and trunks in the wild, gathering moisture from rainfall and dew, and gathering nutrients from particles in the air and falling debris. Plants in this group collect and store nourishment in a cup-shape rosette of leaves called a "tank." Grown on bark slabs or driftwood, tank bromeliads develop enough roots to anchor the plant for support, but the roots absorb little water. Those grown in pots develop a more extensive root system that does absorb water. Refill the tanks of these bromeliads regularly with fresh water and moisten the potting mix when it feels dry.

above This trio of tillandsias includes (clockwise from left) *T. tricolor melanocrater*, *T. stricta*, and *T. caput-medusae*. **opposite, below left** Bromeliads are long-lasting flowering plants ideal for indoor decor.

Provide high humidity to mimic the native conditions. This group includes silver vase plant (*Aechmea fasciata*), guzmania, and blushing bromeliad (*Neoregelia carolinae*). Some tank bromeliads need bright light; others prefer medium light.

Epiphytic bromeliads, or air plants, are often natives to climates where intense evaporation and scarce water challenged their evolution. Some developed scale-covered leaves that absorb moisture; other have cup or vase shapes that hold water. A quick plunge into warm water twice a week keeps air plants hydrated, and a mist of fertilizer solution once a month provides nutrients. Tillandsia is the largest and most varied group of epiphytes. They do not grow well in soil or potting mix. Tillandsias need bright, indirect light.

GROWING BROMELIADS

Half the fun of growing bromeliads is determining imaginative ways to display them. Bromeliads need sufficient light, moisture, and air circulation in order to thrive.

TILLANDSIA
Tack air plants onto a wire wreath form using dabs of hot glue. Take the wreath down for a twice weekly bath in the sink.

SILVER VASE PLANT
A tank bromeliad prefers to be filled with rainwater or distilled water. Hard water can stain the leaves.

EARTH STAR
Grow this terrestrial bromeliad in a pot or dish garden for its rosette of colorful, wavy leaves.

BHG TEST GARDEN TIP

PROMPTING BLOOMS

Easy-to-grow silver vase plant needs bright light to produce flowers. You can help stimulate flowering by placing the plant in a large clear plastic bag with an apple for about two weeks.

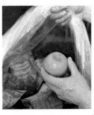

Cactus

Easy-care cacti are intriguing plants for gardeners who look for texture, sculptural beauty, and low maintenance.

Cacti are any busy gardener's dream plants. They thrive in the warm, dry air of many homes and offices. There's no big secret to success with cacti. Their basic requirements are simple: lots of sun, sandy soil, and occasional soakings. It's often said that cacti thrive on neglect.

Desert cacti

Most members of the cactus family are succulents, but not all succulents are cacti. Although cacti and succulents have similar cultural needs, please read more about succulents on page 156. To simplify, this discussion will focus on the best-known desert cacti. These plants store water in their stems, and most have spines.

For the majority of desert cacti, natural conditions consist of cool to cold (40°F–50°F) winters, warm and wet springs, and hot summers. Mimicking these conditions is key to plant success in home gardens. Small desert cacti tend to bloom well indoors as long as they have a distinct dormant period with little or no water and cool or cold conditions. Large desert cacti usually bloom only when mature (20 years or more), and they are prized for their statuesque and spiky forms.

A collection of cacti will help you appreciate their diversity, and there are thousands of species from which to choose. You'll find an array of desert cacti, from squat and round to thin and trailing to flat and candelabra-shape. Some varieties are widely available. Slow-growing barrel cactus (*Echinocactus grusonii*) is round with prominent ribs and works well in a dish garden or an individual pot. Rattail cactus (*Aporocactus flagelliformis*) has long (up to 6 feet) stems covered with delicate spines and is best displayed in a hanging basket. Bunny ears or prickly pear (*Opuntia microdasys*) has broad, flat pads covered with bristly tufts; it stands out in grouped containers.

opposite A wooden frame holds a collection of small desert cacti. A large screw, inserted through the wood, protrudes and reaches into the drainage hole of each pot, holding the pot in place.

HANDLING CACTI

Although cacti are easy to maintain, they merit special handling. Use these tips to avoid a painful poke from the sharp spines that cover most cacti.

Display cacti where they won't easily make contact with skin or eyes, and keep them away from kids and pets who may not be aware of the need to avoid their spines.

Repotting is necessary every few years when a plant pushes out of the pot or the stems reach the edges of the pot. When transplanting cacti, wear heavy gloves and wrap a folded sheet of newspaper around a small cactus; a length of thick fabric around a large one.

If cactus spines stick in your skin, press very sticky tape on the area and then pull it off to remove the spines.

Caring for cacti

Plant cacti in a porous, freely draining mix of sand, perlite, and potting soil. The ideal container is a clay pot, slightly larger than the plant. It must have a drainage hole.

Indoors, cacti like as much sun as possible. Grow them in high light—a south-facing window is best; an unobstructed western exposure is second choice. Give cacti a summer vacation outdoors, setting them in a spot where they'll get some shade.

Water cacti thoroughly but infrequently. Overwatering will rot the roots; let the soil dry between waterings. In the winter, when plants are resting, let an extra week go by between waterings and water just enough to keep the plants from shriveling. Use an all-purpose, water-soluble fertilizer every second or third watering during the growing season.

BHG TEST GARDEN TIP

GROUPING CACTI

A dish garden provides a pleasing way to display a small collection of desert cacti. It includes plants of varying forms, from low and wide to tall and slender. Spread gravel mulch for a neat, natural effect.

Coleus

Colorful, leafy coleus rivals most flowers. This easy-care tropical plant can live indoors with the right care.

As one of the most colorful indoor plants, coleus has had several waves of popularity. When it debuted in Europe as a cultivated plant two centuries ago, the native of Java flaunted its colorful foliage and easy-growing nature. Victorians treasured coleus as a showy bedding plant that overwintered readily as a houseplant grown from cuttings.

The next coleus wave crested in the 1970s when the seed-grown varieties that had changed little since Victorian times moved indoors again as ubiquitous houseplants. These old-school varieties required frequent pinching to remove the negligible flower stalks and to keep the plants growing dense and compact. The ease of coleus propagation from cuttings that originally boosted their popularity, as well as desirable qualities garnered from rediscovered heirlooms, has prompted a recent breeding heyday. Because of the ease of perpetuating coleus and its quick growth, coleus is known as the rabbit of indoor gardening.

Coleus culture

As part of its modern makeover, coleus has been reclassified botanically as *Solenostemon scutellarioides*. The range of plants varies in leaf colors, patterns, sizes, and shapes. Newer varieties bask in bright light or the less sunny conditions preferred by older coleus selections. Many varieties display their richest colors when grown in bright enough light; colors fade in less-than-ideal light. If your coleus becomes spindly or bends toward light, move the plant to a spot where it will get more sun. Pinching off the growing tips, especially when flowers start forming, helps keep a plant bushier.

Coleus do best in temperatures between 60°F and 68°F indoors, with ample humidity. If the air is too dry, you may notice leaves shriveling and falling. Water regularly when the soil surface begins to feel dry, and prevent the plants from wilting.

TRAIN A COLEUS STANDARD

With regular pinching, coleus grows dense with increased branching, and plants can be trained into an impressive standard or lollipop-shape tree within a year or so. Start with a strong single-stem plant; choose a variety that would ordinarily reach about 3 feet. Transplant into a weighty 6- or 8-inch pot to help prevent the plant from becoming top-heavy. Stake the plant to reinforce the main stem. Snip off any side stems from the bottom two thirds of the main stem, leaving the top third to grow bushy. Pinch off the tips of the remaining stems monthly to encourage branching.

Keep coleus going

As a tropical plant, coleus thrives in long, hot summers—better than many gardeners. When moving plants outdoors for the summer, place them in a partly shaded spot and water generously. The plants grow quickly in these conducive conditions and they'll need repotting. Protect them from cold. Take cuttings or prune plants to half their size before taking them indoors in late summer.

To start new plants, snip 4- to 6-inch stem cuttings in late summer. Root the cuttings in a glass of water or in cell packs filled with soilless medium. After roots develop, transplant the young coleus plants into small pots of potting soil, and keep them growing on a sunny windowsill until late spring.

opposite Showy coleus standards (tree forms) happily overwinter indoors given plenty of light. Most varieties of coleus respond well to training as a standard.

Fern

One of the most beautiful foliage plants, frilly ferns are an indoor decorator's dream plant.

The arching fronds of ferns bring simple beauty to indoor settings. Grown in a pot or urn, small ferns add elegance to a dinner table. A large fern in a hanging basket creates a lush, tropical effect. Some species, such as staghorn and tree ferns, form sculptural accents—they truly are living works of natural art.

Helping ferns thrive

The lineage of these primitive plants reaches back more than 300 million years. Earth was wetter then than it is now, but many of the ferns grown today have changed little from their ancient ancestors. The natural habitats of the thousands of fern species vary widely. Delicate feathery ferns (maidenhair, rabbit's foot) favor continually warm and humid tropical conditions; the more leathery-leaf varieties (brake, Japanese holly, bird's nest) tolerate cooler, less humid environments.

Humidity is essential to a fern's well-being. Some varieties need high (60 percent or greater) relative humidity that is difficult to provide in a heated home during winter. Raise the humidity around your plants by setting pots on pebble-filled saucers or trays, or double-pot the ferns, placing the potted fern inside a larger container and filling the space between with pea gravel or sphagnum moss.

When watering ferns, do not let them stand in excess water. Most varieties prefer moist soil. The plants will drop their leaflets if watered too much or too little. Mulching plants with leaf mold, chopped leaves, or chipped bark preserves soil moisture and adds nutrients to soil. Ferns need only half-strength fertilizer every other week during the growing season.

left The delicate or leathery foliage of ferns makes them popular houseplants, and their diversity provides many plants to know and love. Varieties combine well in containers with other plants that also like moist conditions.

Differing needs

Light requirements vary too—from medium to bright indirect light—but no ferns prefer full sun. If a plant grows weakly, move it into brighter light. The light from a north or east window typically sustains a fern. Most of the plants benefit from being moved to a shady spot on the patio for summer, but don't be surprised if they drop leaflets when you move them back indoors. Although ferns generally prefer indoor temperatures between 60°F and 70°F, they'll be less stressed by cooler temperatures than they will be affected by heat—especially dry heat. These plants can also be sensitive to pesticides, tobacco smoke, and leaf-shine products.

FERNS FOR INDOOR GARDENS

When choosing a fern for indoors, see what appeals to you at the garden center, but match the plant's needs to the environment in your home.

1. BOSTON FERN Tolerant of indoor conditions, *Nephrolepis exaltata* develops a large mass of long, arching fronds.

2. CROCODILE FERN An easy-growing fern, *Microsorum musifolium* 'Crocodyllus' has an interesting texture.

3. RABBIT'S FOOT FERN Known for its furry rhizomes, *Phlebodium aureum* requires consistent moisture.

4. ASPARAGUS FERN High humidity is needed by *Asparagus densiflorus*, a member of the lily family.

5. BIRD'S NEST FERN *Asplenium nidus* needs evenly moist soil and high humidity.

6. LEMON BUTTON FERN A small fern with lemon scent, *Nephrolepis cordifolia* grows easily in a terrarium.

7. BRAKE FERN Also known as table fern, *pteris* has differing frond styles, some of which are variegated.

8. STAGHORN FERN Grow *Platycerium bifurcatum* on a bark slab in a humid setting.

9. MAIDENHAIR FERN *Adiantum pedatum* demands humid conditions such as those found in a bathroom.

10. JAPANESE HOLLY FERN Slow-growing *Cyrtomium falcatum* tolerates dry, cold air. Try it in an entryway.

Ivy

Classic vining ivy works as an overflowing container plant or one that can be trained onto a topiary form.

Ivy arrived in America in the form of cuttings that colonists carried with them across the ocean. The plant has proved itself adaptable by thriving in various conditions, outdoors and indoors.

Ivy grows so easily indoors, some people think of it only as a houseplant. It is among the most popular houseplants, with evergreen foliage on wiry, trailing tendrils. Indoors, ivy is most often displayed in hanging baskets. The plants also can be trained to climb a topiary form, trellis, moss pole, or window frame. Ivy even scrambles up rough surfaces such as brick walls because it has aboveground (aerial) roots that allow it to cling.

Types of ivy

Hedera is a small genus with only about a dozen species. The two species commonly grown indoors—*H. helix* (English ivy) and *H. canariensis* (Algerian ivy)—originated in Europe and North Africa, respectively. English ivy, more common as a houseplant, is available in hundreds of varieties with differing shapes and leaf colors.

Collectors describe ivy leaves by their shape: heart, fan, bird's foot, and curly. The basic ivy leaf has three to five lobes in a pleasing hue of glossy green. Leaves range from the largest Algerian ivies to miniature English ivies, including 'Duckfoot', 'Pixie', and 'Lilliput'. Older cultivars produce long stems with widely spaced leaves. Pinching the tips of English ivy encourages branching and a fuller appearance. Many modern cultivars are self-branching and grow more densely on their own.

Green is never monotonous when you invest in ivy. Foliage hues range from yellow-green to eerie blackish green. Some ivies show off a silvery, yellowish, or reddish sheen or contrasting veins; others boast an array of marbled green, gray, gold, and ivory.

left Make an elegant and festive centerpiece with English ivy and a classic urn. Garden centers often have hanging baskets of ivy with long, trailing stems. **opposite, below left** Always a classic pairing: trailing ivy planted in an urn.

Indoor ivy care

Ivies need medium to bright, indirect light and will become spindly in poor light. The plants do best in cool temperatures (60°F–65°F) and like cooler (to 50°F) conditions during winter. Water regularly, allowing the top inch of soil to dry between waterings. Ivy loves a warm shower or bath, especially when dusty, and gleams afterward. When plants take a winter rest, water sparingly but avoid letting the soil dry out completely. During the growing season, feed plants monthly with a high-nitrogen fertilizer. Multiply your ivy collection by taking 4-inch cuttings from the branch tips and rooting them in a soilless medium. Beware: Susceptible individuals can develop a skin rash from handling English ivy.

AVOID YELLOW LEAVES

Overwatering is most often the culprit when ivy leaves turn yellow and fall off, especially during winter. Lighten watering regimen, allowing the top inch of soil to dry between waterings. Watch for tiny webs and reddish spider mites. Rinse them off with a forceful spray of water, or use insecticidal soap to kill them.

MAKE AN IVY TOPIARY

If you love the appearance of topiary but don't have the time or patience for the repeated trimming required by herbs and dwarf evergreens, train an ivy to climb and fill a frame instead.

1 START
Gather materials, including a 10- to 12-inch pot, a 10×24-inch spiral topiary frame, four or five small ivy plants, cloth-covered florist wire, moss, and potting mix. Select plants with long trailers to get a finished effect more quickly. A hefty pot works best to anchor top-heavy designs.

2 SET UP
Press the legs of the topiary frame into the pot full of potting mix. Secure the frame with an added stake.

3 TRANSPLANT
Gently place each plant, tucking the root ball into the soil mix and covering it.

4 TRAIN
Space the plants evenly, giving them room to grow.

5 FINISH
Water the topiary after planting, and thereafter whenever the soil feels dry. Covering the soil surface with decorative moss helps maintain soil moisture.

Winter jasmine dazzles with its marvelously fragrant flowers and vines on a trellis or other support. It is a member of the olive family (Oleaceae), which includes about 200 species.

Jasmine

A plant with so many charms! Jasmine bears beautiful white flowers and emits an intense scent.

Little boosts spirits more than fragrant flowers blooming indoors, especially during the winter months. Jasmine (*Jasminum*) is a favorite aromatic flower. Its scent is smooth, sensuous, and intoxicating, whether the plant is just coming into bud or blossom, or is fully flowering and showering the floor with small, starry-white flowers. When jasmine is placed on a sunny windowsill, the sun's warmth intensifies the perfume, which is neither assertive like a hyacinth nor strident like a paperwhite narcissus.

Choose a jasmine

Not all species of jasmine smell the same; some have no fragrance. They are shrubs or climbing plants that grow well indoors, often blooming in winter, and are even happier when moved outdoors for summer. Vining and twining jasmines require staking and regular pinching to enhance growth and form. Even shrubby varieties have long branches that lean on nearby plants for support.

For vigorous growth and ease of culture, grow J. *polyanthum* (winter or Chinese jasmine). It needs a sunny, cool (40°F–60°F) place indoors throughout fall and into winter for optimum growth and budding. Arabian jasmine (J. *sambac*), such as the leathery-leaved 'Maid of Orleans' and double-flowering 'Belle of India', is shrubbier than most and blooms heavily in summer. Poet's jasmine (J. *grandiflorum*) blooms from spring through fall in temperate regions; it provides an everblooming vine in the South. Other jasmine varieties pique the fancy of collectors. However many plants you have, make the most of jasmine flowers: Toss a handful into a warm bath or float a single blossom in a cup of tea and inhale the fragrance.

What jasmines need

The plants need bright light for abundant flowering. When summering your jasmine outdoors, set it in half-day sun. Weekly watering gives jasmine the moisture it needs without overwatering. Use a half-strength fertilizer monthly throughout the summer. A cool room is key to keeping the plant healthy indoors in winter. Vining plants can reach 6 feet and need annual spring pruning after flowering. Shrubby plants benefit from regular pruning too. Snip the brown or broken stems as well as overly lengthy ones to keep a plant bushy and dense. Prune shrubby plants after flowering, being careful not to remove flower buds. Propagate jasmine by taking stem tip cuttings in early summer.

OTHER FRAGRANT FLOWERING VINES

Vining plants cascading from a hanging basket or clinging to a trellis deliver a soothing sweep of greenery. The fragrance of some flowering vines provides another dimension of beauty as well.

PASSIONFLOWER

Passiflora spp. is another proven indoor performer with fragrant flowers. The vigorous vine will bloom all year given enough light in a south- or west-facing window or under grow-lights.

MADAGASCAR JASMINE

Stephanotis floribunda makes a fine indoor plant with extraordinarily fragrant flowers that bloom in summer. The climbing shrubby vine needs a sturdy support.

WAX PLANT

Hoya spp. are easy-growing vines with beautiful clusters of fragrant star-shape flowers. The older the plant, the more spectacular the blooms.

Orchid

One of the most elegant of indoor flowers, orchids are easier to grow than many people realize.

Do you think you need a greenhouse or other special conditions to pamper these plants that have a prima donna reputation? Definitely not. Bring home an orchid and see how undemanding it can be. If you select one of the easy-to-grow orchid varieties, you'll find they require less care than some common houseplants and will succeed in the environment most home settings provide. Situated near a sunny window, an orchid will surely prove alluring and enjoyable.

Today's orchids

Modern propagation techniques have significantly increased the availability of orchids and lowered the cost of many cultivars. As one of the largest and most sophisticated plant families in the world, orchids astound with their tongue-twisting names. But undemanding *Phalaenopsis* and *Paphiopedilum*—moth and slipper orchids—are among those widely sold at florists, nurseries, grocery stores, and home centers.

When buying an orchid, choose one with grassy green leaves and stems with buds and flowers just starting to open so you can witness the blooming process, which will last a month or longer. With proper care, orchids will grow and rebloom, rewarding you with breathtaking beauty.

Growing orchids indoors

Most orchids grown as houseplants are epiphytes, which are plants that grow naturally on tree branches and have thick aerial roots. Planting an orchid in a pot with lots of side holes and light aerated potting medium allows the roots to absorb plenty of air, which is essential to the plant's survival. Chopped bark, available commercially, is the standard potting medium because it is difficult to overwater. Most orchids' roots will suffocate in soil or soilless potting mix. Sphagnum moss works as a potting medium but proves tricky.

left Grouping orchids on a pebble tray creates a pocket of humid air around them. This group features good options for novice gardeners: cattleya, nun's orchid (*Phaius*), slipper orchid (*Paphiopedilum*), and moth orchid (*Phalaenopsis*).

As an extremely varied group of plants, orchids' other cultural needs differ considerably. They require medium to bright light. Generally, the larger an orchid's leaves, the less light it needs. But if your orchid does not flower, chances are the light level is too low. Average indoor temperatures (65°F–85°F) suit *Phalaenopsis*, *Dendrobium*, and *Oncidium*. Many orchids need a temperature change of 10 to 20 degrees between night and day for several weeks to prompt flowering. Keep plants on the windowsill in a cool room and turn down the thermostat at night to help promote flowering.

Orchids' water needs also vary. But overwatering is the most common way to kill them. Weekly watering works for many commonly grown orchids, but watering schedule will depend on the conditions of your home and the type of pot—water evaporates faster in a dry climate, in a sunny location, and in a clay pot.

DECORATE WITH ORCHIDS

Orchids make delightful holiday decorations. 'Yu Pin Lady' and 'Dong Beauty Queen' *Phalaenopsis* orchids team up in a 6-inch pot staged in a cast-iron tree stand.

ORCHID SUCCESS

Each type of orchid has differing needs. Many orchids come with care instructions, but you can find detailed advice for various types at the American Orchid Society's website: *aos.org*.

WATER Take the pot to a sink and run warm water through the bark medium for a couple minutes. Let the pot drain, then return it to its usual place. If an orchid is left standing in water, the roots will rot.

REPOT Every year or two. Transfer an orchid to a slightly larger pot when its roots outgrow the old one. Replace potting medium with fresh bark, covering bark with a bit of green moss to help hold moisture.

REFRESH When repotting an orchid, snip off roots that are dead, shriveled, or broken. Feed moth orchids year-round; other orchids every other week from spring through fall, using a half-strength formula.

HUMIDIFY Although some orchids can survive low indoor humidity, setting pots on a bed of wet pebbles boosts moisture around them. Plants benefit when they are grouped, which helps humidify the air around them.

Scented Geranium

Although the flowers aren't the main attraction, the scented leaves of this plant are worth a place at the table.

Herb gardeners and collectors have long favored charming plants that invite interaction. Scented geraniums have textured leaves—fuzzy, velvety, or crinkly—that release delicious fragrances of spices, flowers, and fruits when handled. Names reflect the varieties' scents, ranging from lemon to apple, mint, rose, cinnamon, chocolate, and more. These are the scratch-and-sniff plants of indoor gardening. What's more, some are good enough to eat.

Distinctive geraniums

The genus *Pelargonium* includes the flowery darlings grown widely as annuals. Many also have fancy leaves and some varieties known as zonals and regals or Martha Washingtons, grow well indoors. Their scented-leaf cousins grow happily in a sunny garden, but are tender perennials that must be potted and moved indoors over the winter for protection from freezing temperatures.

Scented geraniums grow easily in pots year-round in bright light. Although some varieties have pretty little flowers, the leaves are the most notable feature. Dozens of scented geraniums have a range of leaf shapes and hues, from ruffled to serrated, gray-green to white-variegated. Leaf sizes vary too. Small-leaf varieties, such as lemon 'Crispum', work especially well for topiaries.

left Scented geraniums, such as lacy-leaf variegated mint rose and upright lavender varieties, exhibit diverse characteristics. These plants that invite touching are fun to have around, although their distinctive scents may be left to the nose of their beholder.

Indoor gardening advice

Grow scented geraniums in a south- or west-facing window that offers bright light. If you give lower light for the plants to overwinter in a semidormant mode, provide cooler temperatures between 50°F and 60°F.

Pinch or snip stem tips regularly to keep the plants growing strong and bushy. To generate new geraniums easily, take 4-inch-long stem cuttings in fall. Overwinter them on a sunny windowsill and you'll have substantial new plants ready to move into larger pots in late spring. Let them grow outdoors for the summer, if you like. They'll likely need repotting and cutting back by a third when fall arrives.

Water scented geraniums thoroughly, then let the soil begin to dry before watering again. Soil that is too wet causes root rot. Keep plants healthy by watering with a half-strength all-purpose fertilizer solution throughout the summer. Lay off feeding during the winter.

Using scented geraniums

Dried leaves add distinctive scents to potpourris and sachets; citrus, spice, and rose fragrances blend especially well with rose petals and lavender flowers.

Cut fresh stems and tuck them into bouquets—they'll provide a long-lasting scent. Enjoy the leaves' refreshing flavors by using them in iced tea and baked goods.

SCENTED GERANIUM VARIETIES

Scented geraniums are distinguished by their fragrance. The sizes of the plants and their leaves vary, as do other leaf characteristics.

1. ATTAR OF ROSES A reliable rose-scented variety, *P. capitatum* may bear lavender-pink flowers if conditions are bright enough.

2. CHOCOLATE MINT *P. tomentosum* has a dark variegation at the center of the leaves and a minty (but chocolaty) fragrance.

3. OLD SPICE With its spicy fragrance and compact habit, *P. × fragrans* is a good choice for topiary.

4. SPANISH LAVENDER One of the larger species, *P. cuculatum* reaches 2 to 3 feet tall and has velvety, almost-cupped leaves.

5. FROSTED Only a few scented geraniums have variegated leaves. *P. citrosum* 'Frosted' has white tips.

6. LEMON An excellent culinary variety, *P. crispum* has small crinkled leaves with a strong lemon flavor.

7. COCONUT A round-leaf variety, *P. grossularioides* has a sweet, light fragrance. It is lovely in bouquets.

8. CITROSA Sold as a mosquito-repelling "citronella plant," lemon-scented geranium is unproven as an insect deterrent.

9. PINE With its deeply cut leaves, *P. denticulatum* has a pungent scent of pine.

10. PEPPERMINT A refreshingly pungent variety, *P. tomentosum* has large velvety leaves. Its trailing habit suits hanging baskets.

Succulents

Easy-care succulents are available in a wide range of leaf shapes, colors, and textures, and are wildly popular.

Fleshy, water-storing leaves, stems, and roots earn these plants the catchall classification of succulents. Especially well-adapted to arid climates, succulents hail from worldwide locales, ranging from sunny and dry to damp and cloudy.

Many succulents form simple rosettes. Some sprawl along the soil's surface and some grow upright. Others tumble over pot edges, leaf-studded stems cascading stiffly. Leaf colors vary widely, from chartreuse to gray-green, reds, and almost black. Sometimes flowers appear in spring or summer, depending on the variety, and especially when plants grow outside in full sun. Planted solo or in artful combinations, these jewels weave textural tapestries indoors as well as outdoors, and they look good year-round.

Succulent culture

The laid-back personalities of succulents make them generally easy to grow and a good option for gardeners of any skill level. They're naturals for containers. It's fun to choose from the array of plant varieties then display the succulent in a complementary pot that highlights plant form and color. Displays can be simple and geometric, or combinations of plants in a texture-rich dish garden.

Many succulents' diminutive size, from 2 to 6 inches, is another reason for their popularity in indoor gardens. Small pots of them fit as readily on a windowsill as in a wall planter. Tuck their shallow roots into the pockets of a strawberry jar or into a wooden box. Clay and other porous containers make it easy to control the moisture level of the potting medium because they allow evaporation. (Plastic, glazed ceramic, and glass containers hold moisture longer.) Repot plants every few years. Succulents prefer to be crowded in a container, so wait until the roots have no more growing room before repotting.

left Small flower-shape succulents (*Echeveria, Sedum, Pachyveria* and silver squill) reach elevated heights in mercury glass vases. Bits of reindeer moss complete the decorative effect.

Excellent drainage and bright light are essential to the well-being of succulents. Their roots do not like standing in water, and plants will become thin, elongated, or pale without adequate light. Add sand, pea gravel, or perlite to potting mix for a quick-draining blend. Place succulents in the brightest light available indoors during the winter months.

The surest path to premature death of succulents is not through frosty nights on a windowsill but rather by overwatering. Water plants sparingly through the winter—just enough to keep roots alive and prevent plants from dehydrating and shriveling. In summer, water more often, especially if you move plants outdoors. Feed plants only during the growing season, using a quarter-strength solution.

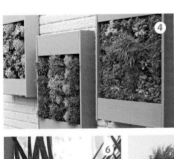

SUCCULENT PLANTING IDEAS

Many succulents are stiff and bulky; others are loose and flexible. Whatever their form, the plants have a sculptural presence and provide textural materials for decorative displays.

1. AEONIUM Tall containers hold top-heavy 'Sunburst' aeonium, burro's-tail (*Sedum morganianum*), and echeveria.

2. JADE PLANT For interesting results, put a vintage treasure to work when displaying a jade plant (*Crassula ovata*).

3. BURRO'S-TAIL A vintage beverage bottle crate becomes a vertical garden planted with burro's-tail and ghost plant (*Graptopetalum*).

4. SEDUM Creeping and low-growing sedums create artistic mosaics in hanging wall planters.

5. SEMPERVIVUM When planting a repurposed container, improve drainage by poking a hole in the bottom of each planting pocket.

6. ALOE In a sunny kitchen window, Aloe vera is pretty and practical. Gel inside the leaves provides first aid for burns.

7. ECHEVERIA Wrap the root balls of echeveria with sphagnum moss before wiring them to an evergreen wreath with a heavy-duty wire base.

8. AGAVE Stone mulch and a neutral-color pot showcase an agave or other subtle-hued succulent.

9. STRING OF PEARLS *Senecio rowleyanus* provides the jewelry for a mixed planting of variegated aeonium and sticks on fire (*Euphorbia tirucalli*).

10. KALANCHOE This panda plant (*Kalanchoe tomentosa*) lives in a copper pot. Embellishments include princess pine, variegated boxwood, and eucalyptus.

CHAPTER NINE

Houseplant Encyclopedia

Possibilities for indoor gardens are better than ever, from traditional plants to newest varieties. Start here to find the options best suited for your home.

YOU
— SHOULD KNOW —
Some alocasia varieties
with enormous leaves are
especially popular in outdoor
container gardens. The tubers
must be stored in a cool,
dry, frost-free location
over winter.

Alocasia
(Alocasia hybrids)

The striking foliage of this tropical inspired its common name, elephant's ears. Only a few of the dozens of species and hybrids are sold as houseplants. The green, blackish-green, or silver-green foliage usually has contrasting veins and lobed or wavy margins. The plants reach 2 to 6 feet tall.

Best site

This showy houseplant does best with bright to medium light in winter; indirect, medium light in summer—never direct sun. Alocasia depends on warm (65°F–70°F) temperatures and high humidity. A pebble tray is effective in raising humidity around a plant. Avoid exposing the plant to temperatures below 50°F, cold drafts, and extreme temperature changes. The plant benefits from living outdoors in a mostly shaded spot during the summer, especially if the area where you live is warm and humid.

Growing

Use a well-draining, moisture-control potting mix. A 2-inch layer of bark mulch helps keep the potting mix evenly moist. Water as needed to maintain soil moisture, but avoid creating soggy soil. Allow the soil to become a bit drier during the winter. Alocasia is sensitive to the chemicals in municipal water, so use room-temperature rainwater whenever possible to hydrate your plant. Feed the plant monthly during the growing season, using a standard liquid fertilizer for foliage plants; do not feed in winter. Bathe the plant or wipe its satiny leaves with a damp cloth occasionally to remove dust and grime. Propagate alocasia by carefully removing side shoots with roots attached; or by lifting the plant from the pot, removing a rhizome or tuber, and cutting it into pieces for rooting. This plant is generally pest-free, but you may notice spider mites, mealybugs, or scale on a stressed plant.

Special notes

Alocasia can be temperamental about its growing conditions. When not warm and humid enough, the plant tends to drop leaves. Too much moisture causes the plant to rot. Leave the top of the rhizome or tuber exposed when repotting to help prevent rot.

1

ALOCASIA VARIETIES

1. 'MOROCCO' is a fast grower with pretty pink stems and glossy dark green leaves with bright greenish-white veins and burgundy undersides.

2. 'POLLY', a dwarf cultivar, grows to 2 feet tall and has greenish-white veins.

3. 'STINGRAY' leaves resemble a sea creature with a long, pointed tail. Plants grow 5 to 6 feet tall.

2 3

Aluminum Plant

(Pilea spp.)

More than 600 species belong to this genus. Those best suited to indoor gardening are 6 to 12 inches tall with foliage of various colors and textures. They're attractive enough to stand alone, but contrasting species work beautifully in groupings. The plants also have appeal because they grow so easily.

Best site

Provide this tropical plant with medium to bright, indirect light in an east- or west-facing window. Overly strong light will burn the leaves; too little light results in spindly growth. The plants prefer a warm to high temperature (60°F–70°F). Cold air and cold water damage the leaves. Pileas will not thrive if exposed to cold drafts, drying winds, or low humidity.

Growing

The dark green leaves of aluminum plant (*P. cadierei*) appear to be painted with streaks of silver. The leaves of other species are splashed in copper or red and may have a quilted appearance. Some grow upright and are adapted to terrarium culture; others creep or trail, making them useful in hanging baskets. Let the soil begin to dry out slightly between waterings, but not so dry that the plants wilt or become droopy. Reduce watering during the winter. The plants have small root systems and rarely need repotting. Good drainage is essential. Fertilize plants regularly from spring to fall, then lay off the plant food through the winter. Aluminum plant and its relatives are most attractive as young plants. Pinching or trimming the stem tips periodically helps keep a plant lush and bushy. Start new plants from stem cuttings—they grow easily.

Special notes

Sometimes aluminum plant is considered short-lived. But if you pinch off the stem tips and start new plants from the cuttings, you'll be rewarded with new branching on older plants as well as vigorous young ones. When humidity is low, spider mites may develop, causing the leaf edges to brown. Mealybugs and scale may also appear. Artillery plant can become a nuisance by sending its seeds flying into the soil of nearby plants; prevent overpopulation by keeping it isolated.

YOU
— SHOULD KNOW —
Creeping pilea (*P. nummulariifolia*), is also known as creeping Charlie, not to be confused with the lawn weed also called creeping Charlie (*Glechoma hederacea*).

ALUMINUM PLANT VARIETIES

1. ALUMINUM PLANT (*P. cadierei*) leaves are splashed with silvery markings.

2. 'MOON VALLEY' (*P. involucrata*) has deeply corrugated and dramatically colored leaves.

3. 'NORFOLK' (*P. spruceana*) is distinguished by bronze-green foliage with silver bands. The oval leaves have a quilted surface.

4. CREEPING PILEA (*P. nummularifolia*) has bristly, corrugated leaves on a vining plant.

5. ARTILLERY PLANT (*P. microphylla*) has tiny leaves on a branching, bushy plant.

YOU
— SHOULD KNOW —
Like true aralias, but not related to them, false aralia (*Schefflera elegantissima*) has interesting, textural foliage and can grow to 8 feet tall.

Aralia
(*Polyscias* spp.)

This group of about 40 species of trees and woody shrubs has varied appearances but similar needs. Most have interesting foliage and stem textures. Aralias have a reputation for being difficult to grow indoors, but the plants can be successful and rewarding if you provide the right environmental conditions.

Best site

Provide aralias with medium to bright, indirect light. Low light causes spindly growth. Aralias insist on warm temperatures (60°F–85°F) and high humidity (65 percent). Without these conditions, the plants will drop their leaves. Placing an aralia on a bed of wet pebbles helps boost humidity. Keep the plants away from cold drafts and avoid letting the temperature drop below 50°F at night.

Growing

Let the potting mix dry slightly between waterings. Water the plant less during the winter. Aralias grow somewhat slowly and don't often need repotting. But a plant will let you know when it needs a larger pot: It will no longer take up water. During the growing season, fertilize every three weeks, using an all-purpose plant food. Prune as needed to maintain an aralia's overall size and shape. If your plant drops many of its leaves over the winter, it will start to fill in during the spring if you trim the plant a bit and set it in a place where it will receive more light.

Special notes

Aralias grow to 8 feet tall and 3 feet wide indoors. Some types can be kept smaller with regular pruning. They can even be trained into beautiful bonsai. Leaf drop and leaf-edge browning are signs of inadequate humidity. Although aralias require high humidity, they are sensitive to overwatering. At first, overwatering causes shiny dark spots on the backs of the leaves, which will then turn yellow and fall off. The plant may wilt if it is kept too wet. Aralias are also prone to root rot from overwatering. They are not usually prone to insect invasion, but spider mites, mealybugs, and scale insects may become problems. Treat them with horticultural oil. Aralias are poisonous. Children or pets can become ill if they chew or eat aralia leaves.

ARALIA VARIETIES

1. MING ARALIA (*P. fruticosa*) has feathery leaves and a textured trunk. It develops a treelike form, even at a young age. Group several in one pot to create a shrubby plant.

2. BALFOUR ARALIA (*P. scutellaria* 'Balfourii') is a woody plant with glossy, scalloped leaves.

3. VARIEGATED BALFOUR ARALIA (*P. scutellaria* 'Balfourii Variegata') offers round, deeply veined, variegated leaves. It grows to 8 feet tall.

Arrowhead Plant

(Syngonium podophyllum)

Also called arrowhead vine, this tropical trailing plant grows easily indoors, whether in its own pot or in a mixed planting. As the plant matures, its leaves become more deeply lobed and some of the compact stems change to vines. The leaves may be deep green, white-variegated, or tinged metallic pink.

Best site

The native region of arrowhead plant ranges from Mexico to Central America. The popularity of this plant has propelled intensive hybridizing, and now more compact forms of *Syngonium* are available. Newer varieties also boast unusual coloring (bronze or pink hues) and increased disease resistance. Arrowhead plant will thrive, remaining compact and showing off its best color, in medium to bright, indirect light. This attractive foliage plant tolerates various indoor conditions, with the exception of direct sun and very low light. Too much sunlight will burn or bleach the leaves; too little light prompts spindly, weak, and pale growth. Most humidity levels and average temperatures suffice, as long as the plant is not exposed to cold drafts. Temperatures of 60°F to 75°F are ideal.

Growing

When arrowhead plant matures and develops its vining characteristics, it can reach from 4 to 10 feet tall. A juvenile plant and one kept from vining by pruning will be about 1 foot tall. Repot your arrowhead plant each spring in fresh potting mix enriched with compost. Fertilize monthly from spring through fall. When watering, aim for even moisture—not too wet and not too dry. It won't hurt if the soil starts to dry a little between waterings. Avoid overwatering. The plant will rot in soggy soil. Propagate an arrowhead plant from cuttings of pencil-size stem ends.

Special notes

Arrowhead plant is an excellent candidate for a hanging planter or a pot placed on a pedestal. Give an older plant grown in a pot and upright support—a moss-filled pole works best—to help it climb, or snip off the vining stems to keep the plant more compact. When training a plant to a moss pole, pin the stems with stubby aerial roots to the pole and keep the moss damp. A monthly shower helps prevent spider mites. Also watch for scale and mealybugs.

YOU
_ SHOULD KNOW _
Arrowhead plant is toxic if chewed or eaten. Keep it away from kids and pets.

1

2

3

ARROWHEAD PLANT VARIETIES

1. 'PINK ALLUSION' is one of the decorative Allusion Series characterized by pink veining or pink-blushed leaves.

2. 'LEMON LIME' is among dozens of newer cultivars that has vivid lime-green leaves marked by white veins that fade to a whiter leaf over time.

3. 'HOLIDAY' stands out with its ornamental foliage distinguished by berry-pink veins.

Baby's Tears
(*Soleirolia soleirolii*)

A delicate, creeping plant with tiny foliage and fine stems, baby's tears grows well in a shallow pot. The 2-inch-tall groundcover spreads quickly in a terrarium. Plant it in a hanging container, where it will trail over the pot's edges or as living mulch in the pot of a larger plant.

Best site

Baby's tears does best in bright, indirect light but it will also grow in medium light. Plants prefer temperatures on the cool side. Provide average room temperatures from 60°F to 75°F. High humidity is essential. Baby's tears also prefer good air circulation, so when planting in a terrarium, allow for some air flow by leaving the top open.

Growing

Add coarse sand to the potting mix to boost drainage (plants were originally found in Corsica and need well-draining soil). Keep the potting medium evenly damp, but not soggy. Err on the moist side because allowing the soil to dry out damages the plant's roots and causes dieback. If too soggy, however, the roots can rot. Baby's tears grows close to the ground and can be used in the garden; pair with ferns and other plants that like indirect light and lots of moisture.

Special notes

Baby's tears flowers in late spring; blooms are small and white. To keep baby's tears in a mound, simply pinch back stems to help it keep its shape. Otherwise it will drape nicely over the edges of pots. Gently divide or prune the plant periodically to keep it under control. If the plant dies back due to less-than-ideal conditions, trim off the damaged portion. Baby's tears will regenerate from the remaining healthy roots. Baby's tears is an easy rooter; just place a piece of stem into a moist soil mix to propagate. In some climates, baby's tears can be considered a weed because of its ease of rooting and its wandering ways.

BABY'S TEARS VARIETIES

1. GREEN is the most common leaf color of baby's tears. Plants come in varying shades of green, from light to dark.

2. 'AURORA' bears delicate, tiny yellow leaves. Plants spread quickly, forming a golden mat of foliage.

Bloodleaf
(Iresine herbstii)

This ornamental plant gets its common name from the intensely colored foliage. It's also called beefsteak plant. Given adequate light, the dark purple leaves of red bloodleaf have crimson veins and pink stems. Other varieties have different coloring, such as lime-green leaves with yellow veining. This native of Brazil grows upright and reaches 2 feet tall.

Best site
Set the plant in bright, indirect light. Provide average warmth (60°F–75°F) and cooler (not below 55°F) conditions in winter, with average to high humidity (above 30 percent).

Growing
During the growing season, keep the soil evenly damp and fertilize monthly. In winter, cut back on watering to prevent root rot. Snip off the stem tips often to promote dense growth. Propagate by stem cuttings. The leaves of bloodleaf fade and become dull if the light is not bright enough. An east- or west-facing window provides the bright indirect light this plant prefers.

Special notes
Leaves tend to fade if the light is not bright enough. The light must be indirect or filtered, or the leaves will bleach out and the edges will brown. Pair red bloodleaf with yellow-and-green striped 'Song of India' dracaena for a beautiful combination.

YOU _SHOULD KNOW_
This showy houseplant can also be used in containers outdoors. Both red and green types make beautiful additions to pots with flowers or other foliage.

BLOODLEAF VARIETIES AND DETAILS

1. BLOODLEAF is a red variety that offers small deep-red flowers with lighter red veins on plants that can grow 4 feet tall.

2. PINCHING BACK new leaves will encourage more shrubby growth of plants.

3. 'BLAZIN LIME' bears rich green foliage and bright golden veins on a 3-foot-tall plant.

YOU
SHOULD KNOW
The leaves of cast-iron plant add sleek greenery to floral arrangements. Cutting stems at the base also prompts new growth.

Cast-Iron Plant

(Aspidistra elatior)

Aptly named for its durable constitution, this plain but popular houseplant grows in almost any condition. Cast-iron plant has leathery oblong leaves that reach about 2 feet tall. Typically dark green, some plants also have variegated leaves. The plant grows slowly, and it can remain in the same pot for years.

Best site

Thanks to its tough-as-nails nature, this plant will survive low light and extremes of humidity. But it responds to good care and medium to bright, indirect light by producing shiny, healthy leaves. It's a good choice for a home or office where a spot of green is needed, but natural light is limited. The plant responds well to artificial lighting. Cast-iron plant is not at all fussy about temperature or humidity, but it will be most content between 50°F and 80°F.

Growing

Allow the soil to dry somewhat between waterings. Cast-iron plant tolerates occasional missed watering, but it won't cope well with overwatering. The plant grows from a rhizome that can rot if overly saturated. Unlike most plants, it resents being transplanted. Because cast-iron plant grows so slowly, start with a large plant if you want its full effect. Unless you start with a large plant, wait for the plant to fill its pot before repotting. When repotting, also divide the rhizome if you want more plants. Each division should have at least three leaves attached to it. A monthly shower will rinse dust off the leaves. Snip off any yellowed leaves. Fertilize every three or four months with half-strength plant food for foliage plants. Fertilize variegated plants less often to maintain their variegation.

Special notes

The plant will rot in soggy soil. If fungal leaf spots become a problem, cut out the damaged leaves and cut back on watering. Spider mites can proliferate on the undersides of a cast-iron plant's leaves in dry situations. Mealybugs can also be problematic.

CAST-IRON PLANT VARIETIES

1. 'MARY SIZEMORE' is a dwarf plant that grows to 18 inches tall and has dark emerald green leaves with white spots.

2. 'ASAHI' shows off 20-inch-long pale green leaves. The upper third of the leaves develop an airbrushed white variegation as they mature.

Chinese Evergreen
(*Aglaonema* spp.)

One of the most durable and attractive foliage plants, Chinese evergreen grows upright with nonbranching stems and leathery spear-shape leaves. Foliage color and size differ widely by variety. This slow-growing plant can cope with low-light conditions. Newer varieties sport red- or pink-variegated foliage.

Best site

An especially pleasing green plant for poorly lit places, Chinese evergreen's silver-, white-, and yellow-variegated forms show off their best coloring when grown in medium light. This native of South Vietnam and the Philippines prefers warm temperatures (60°F–80°F) and high humidity. Protect it from cold drafts and drying winds.

Growing

Place several plants in a shallow and wide pot for a bushy appearance. Repotting is seldom needed and the plant likes to be pot-bound. Repot only when the roots have filled the pot. Allow the soil to dry slightly because the roots are prone to rotting. Shower the plant monthly with warm water to clean dusty foliage. If you grow Chinese evergreen in low light, fertilize the plant only once or twice a year, using an all-purpose plant food. A plant grown in brighter light should be fed monthly in summer. Propagate Chinese evergreen by dividing the root ball: Cut the root ball in half and pot each division. Chinese evergreen blooms occasionally indoors, but the flowers are insignificant. Long-lasting, bright red berries follow. You can sow the seeds from the berries to propagate new plants.

Special notes

Chinese evergreen leaf edges turn brown in cold or dry air or if mineral salts build up from fertilizing or watering. An older plant may lose its lower leaves over time, exposing a short, bare woody stem. To hide this stem and revitalize the plant, cut off the bottom of the root ball and then repot the plant, covering the bare base of the stems. New roots will grow from the buried portion. Chinese evergreen sap contains oxalic acid, which can irritate the skin, mouth, tongue, and throat if chewed.

YOU
_ SHOULD KNOW _
In parts of Asia, it is considered good luck to grow a Chinese evergreen. In Thailand, the plants are called "smiles of fortune plants."

CHINESE EVERGREEN VARIETIES

1. 'SILVER QUEEN' is characterized by lance-shape leaves with silver markings. This compact grower reaches 1 to 2 feet tall.

2. 'WHITE LANCE' has especially slender leaf blades that are mostly silver with a green margin. It grows densely up to 42 inches tall.

3. 'VALENTINE' wins hearts with its especially colorful pink-splashed, dark green foliage. This slow-growing cultivar was hybridized in Thailand.

4. 'EMERALD BAY' is one of the easy-to-grow Bay Series. It sports silvery-gray leaves that are variegated green along the leaf perimeter.

5. 'SIAM AURORA' is a hybrid from Thailand. It has stunning magenta stems and leaf edges contrasted with chartreuse green leaf blades.

YOU
— SHOULD KNOW —
Clivia needs well-draining soil. Amend the potting mix with compost, perlite, and grit. Repot only when the plant begins pushing out of the pot.

Clivia
(Clivia miniata)

Also known as Kaffir lily, this exotic cousin of amaryllis grows from a crown with thick roots. Known for its strappy, glossy evergreen leaves, clivia produces beautiful clusters of trumpet-shape flowers on stiff stalks from late winter through early summer. The plant needs a winter rest to ensure rebloom.

Best site
The bright, indirect light from an east window suits clivia. Throughout fall and into winter, clivia must rest in a cool (40°F–50°F) room to prepare for its next bloom cycle. Move clivia outdoors for the summer; place in a protected spot. Before the first freeze, move your plant indoors.

Growing
Plant top-heavy clivia in a pot large enough to keep it from tipping over. You can also stake plants. Crowded roots left undisturbed produce the best blooms. Let soil dry between waterings. Water just enough during the rest period to keep leaves hydrated. Fertilize monthly during spring and summer.

Special notes
Give the plant plenty of room to grow. In the landscape, clivia forms a large, spreading clump; it must grow in frost-free locations. In a container, new offsets need room to develop, so choose a sizable pot for your plant. You can replant the offsets to make new plants. You can also propagate clivia by dividing the clump. You can also collect seeds from clivia to grow. Seeds germinate best in spring, but require a little patience: The plants can take 4 or so years to bloom from seed. To keep clivia looking its best, peel off the lower leaves as they turn yellow and shrivel. When the flowers are done blooming, clip off the spent flower stalks.

CLIVIA VARIETIES

1. ORANGE is a common flower color of clivia. This flower has a yellow throat.

2. YELLOW clivia can range in color from soft yellow to deep yellow.

Coffee Plant

(Coffea arabica)

Coffee plant is a dark green shrub that grows to 6 feet tall. Less demanding than many other exotic plants, coffee plant matures in five years or so, and may produce fragrant flowers and red berries or beans. It has potential as a long-lived and entertaining plant.

Best site

Coffee plant grows best in the bright light of a south-facing window, but prefers some shade in summer. Temperatures between 65°F and 75°F, and high humidity (65 percent or higher) are ideal. Keep the plant out of cold drafts and wind.

Growing

Water the plant generously throughout spring and summer to keep the soil evenly damp. Ease off watering slightly in fall and winter. Coffee plant doesn't like its soil too wet, so make sure the container it is growing in has adequate drainage. Feed every two weeks during the growing season with half-strength fertilizer formulated for acid-loving plants. You can take your coffee plant outdoors for the summer; place it in a shaded location. Make sure to bring the plant back indoors before frost.

Special notes

Pinch off the growing tips of coffee plant periodically to encourage shrubby growth. Plants can grow up to 6 feet tall indoors, so to keep the plant at a certain height, pinch off the top new growth. The best time to prune coffee plant is spring. Keep an eye out for insect pests such as spider mites and mealybugs. Transplant coffee plant only when the roots have filled the pot. Coffee plants will produce fragrant white flowers which will develop into berries; in order for the flowers to develop into berries, plants must be hand pollinated.

YOU SHOULD KNOW
A coffee plant may not flower and fruit indoors, but the dark, wavy foliage is reason enough to grow the evergreen shrub.

COFFEE PLANT DETAILS

1. PINCHING off new growth will help plants grow shubbier.

2. COFFEE BEANS are the fruit of the coffee plant, and what the beverage coffee is made from. In warm climates, this plant can be grown outdoors and will produce flowers and fruit. As a houseplant, fruiting most likely will not occur.

YOU
— SHOULD KNOW —
Croton is poisonous and should be kept away from children and pets.

Croton
(Codiaeum variegatum pictum)

Croton is valued for its vivid foliage. Leathery and stiff, the leaves vary in size, shape, and variegation. The red, yellow, and green color markings vary among leaves on the same plant. Crotons grow upright and are often potted in multiples to create a bushier form.

Best site

Croton does best in bright light, especially during winter. Given less than bright light, the plant's leaf colors fade and lower leaves drop. Young plants can adapt to lower light and drier air. Once settled, the plants reach full height (typically 4 feet). Many crotons sold throughout North America are taken from a sunny southern state, packed in trucks, and shipped to stores where they're immediately put on sale. But mature plants can react badly if moved from their usual growing location, losing leaves en masse and even dying. If your home has an environment unlike Florida's, start with a small plant. If you buy a full-size specimen, ask for a well-acclimated plant that has been in a local greenhouse for at least two months. Croton requires temperatures above 60°F and high humidity (65 percent). Protect the plant from cold drafts. If the leaf edges turn brown, the temperature may be too low.

Growing

Keep the potting mix damp—neither soggy nor dry. If a croton's leaf tips turn brown, the potting mix is not damp enough or the humidity is too low. Fertilize croton monthly during the growing season with all-purpose plant food; stop feeding over the winter. Propagate croton from stem cuttings in early summer, removing all the leaves except the top two from the stem. Crotons can also be propagated by air layering. This plant benefits from summering outdoors.

Special notes

Showering the plant monthly keeps the leaves shiny and dust-free and helps prevent spider mites. Watch for mealybugs and scale. Pinch off the stem ends occasionally to stimulate branching. Croton doesn't often need pruning, but you can cut it back by one third to one half to revitalize the plant, if necessary. The plant's sap stains, so be cautious when pruning.

CROTON VARIETIES

1. PETRA CROTON, otherwise known as autumn plant, hails from India, Malaysia, and some South Pacific Islands.

2. 'MAMMEY', or fire croton, is known for its elongated, slender, and slightly spiraled leaves.

3. 'GOLD DUST' features golden-yellow speckles all over the lance-shape leaves and grows to 3 feet tall.

4. 'TETRA' is a newer cultivar that boasts brilliant markings. As the young green leaves mature, they develop bright hues of pink and yellow.

5. 'MANGO' sports top leaves that are bright yellow, whereas the bottom leaves range from dark orange to almost black.

Dieffenbachia
(*Dieffenbachia* hybrids)

Native to Central and South America, this striking foliage plant has an architectural presence. Also called dumb cane, it grows 4 to 6 feet tall, and can be planted in a group for a shrublike effect, or singly and treelike. Ingested sap from the plant stem causes temporary speechlessness and much pain, hence its common name.

Best site
Grow dieffenbachia in medium light. Inadequate light causes the plant to lose its lower leaves. Ideal conditions include average humidity (30–65 percent) and warm temperatures from 80°F during the day to 60°F at night. Keep the plant out of cold drafts.

Growing
Plant several young dieffenbachias in one pot to disguise their canes or stems and create a shrubbier appearance. As a dieffenbachia grows, it sends up the thick canes from which the variegated leaves unfurl. It naturally drops its lower leaves as it ages. Mature plants may eventually reach ceiling height. They can be cut back to stimulate new growth and can be propagated by stem cutting or air layering. Cut the plant back to its base when it becomes overly leggy. Use the upper, leafy portion as a cutting and let the stump regenerate. A new plant will also develop from a leafless 3-inch-long section of stem laid on its side and half-covered with potting mix. Wear gloves and handle cuttings with care to avoid ingesting the sap. Allow the potting mix to dry slightly before watering, but be aware that allowing the plant to wilt can cause severe leaf drop. Fertilize only monthly during the summer, using a diluted plant food. Wipe the leaves with a damp cloth every other month to remove dust. Remove faded or damaged leaves.

Special notes
Poor drainage and standing water cause root rot. If the conditions are too dry, the leaf tips turn brown. Leaves develop a washed-out appearance when the plant is exposed to excessively bright light. Spider mites, aphids, mealybugs, and scale are possible pests. This is not a good plant to have in the same household with young children or chewing pets. But it is an outstanding choice for indoor displays, starring independently or adding tropical appeal to a grouping.

**YOU
— SHOULD KNOW —**
Double-potting a dieffenbachia helps prevent a top-heavy plant from tipping over. Set the plant's pot inside a larger heavy one.

DIEFFENBACHIA VARIETIES

1. 'TROPIC MARIANNE' is a hybrid in the D. amoena Tropic Series. It has nearly variegated leaves that are very light.

2. 'DELILAH' has especially thick silvery leaves and the ability to survive in lower humidity conditions that are typical indoors.

3. 'STAR BRIGHT' is known for its narrow (half the width of most varieties), slightly arching leaves with bright creamy white markings.

4. 'CAMILLE' is a compact plant that grows 2 to 4 feet tall. The leaves have creamy variegation with a narrow band of green on the leaf edge.

5. 'TROPIC SNOW' is another hybrid in the Tropic Series. It has pale green and cream variegation on extra-large leaves with a lot of substance.

Dracaena

(*Dracaena* spp.)

This large group of plants is extremely popular for its wide variety. Many dracaenas have tall, thick stems topped with tufts of narrow swordlike leaves. Some grow into large treelike plants. They may reach 10 feet or even taller and serve as imposing pieces of living artwork indoors. Other varieties grow shrubbier and have wider, arching leaves. Most dracaenas are chosen for their foliage traits, whether multicolored, striped, speckled, or banded with contrasting margins.

Best site

Medium light suits most dracaenas, but the plants are typically sturdy and adaptable. Most varieties can tolerate lower light, but their growth will be very slow, and variegated ones can lose the brightness of their coloring. Bright, indirect light in winter pleases the Madagascar dragon tree (*D. marginata*) but direct, full sun is too much for all dracaenas grown indoors. During the growing season, dracaenas grow best in average temperatures between 60°F and 80°F. In the winter, the plants adjust to cooler conditions (55°F).

Most varieties need average humidity (35–65 percent) and they benefit from the increased humidity that comes from grouping plants. Here again, Madagascar dragon tree is an exception—it can grow well in drier conditions, but extremely low humidity or too-dry potting mix can prompt the plant to shed its lower leaves. If humidity is too low, dracaena leaves commonly develop brown tips and yellow margins. The leaves may become soft and curled with brown edges if the plant is situated too close to a cold, drafty window during the winter.

Growing

In general, dracaenas grow slowly. They remain attractive for many years, and their leaves are long-lived until they lose most of the lower leaves. When that happens, start the plants over again from cuttings or air-layer them. Taller and tree-form varieties must be cut back occasionally, such as the corn plant (*D. fragrans*), which can reach 20 feet. 'Song of India', the best known pleomele (*D. reflexa*) tends

to grow at odd angles and must be pruned to keep it under control. It can become tree-size (4 to 8 feet) or trimmed within bounds. Madagascar dragon tree also eventually grows too tall for indoor use and must be cut back or rerooted. Most dracaenas need repotting periodically to give the roots more growing room.

Allow the top inch of potting mix to dry before watering dracaena. Pleomele benefits from more evenly damp soil than other dracaenas. Lucky bamboo (*D. sanderiana*) is often grown in water, held up by decorative stones—clearly, it would also appreciate moist soil. Reduce watering of all varieties somewhat during the winter, but do not let the potting mix dry completely. Feed dracaenas using half-strength all-purpose fertilizer every other month during the growing season. Do not fertilize when growing the plants in low light. Excess fluoride and mineral salts buildup cause brown leaf tips, so use rainwater whenever possible.

Special notes

Spider mites can be a problem in low humidity, but a monthly shower or wiping the plant's leaves with a damp cloth will control the pests. Mealybugs and scale are less common problems. Over time, dracaenas lose their lower leaves and can become gangly. Grouping plants of different sizes helps minimize their bare stems.

CANELIKE VARIETIES

1. MADAGASCAR DRAGON TREE (*D. marginata*), or red-edge dracaena, often has leaves with lengthwise stripes and bent or twisted stems.

2. 'MASSANGEANA' (*D. fragrans*), also called variegated corn plant, occasionally produces sweetly scented blossoms indoors, given ideal conditions.

3. 'ART' (*D. deremensis*) bears straplike green leaves with creamy yellow margins.

4. 'LIMELIGHT' (*D. deremensis*) produces lime-green leaves on tall, canelike stems.

5. 'LEMON LIME' (*D. deremensis*) has striped green, gray, and white foliage with chartreuse leaf margins.

SHRUBBY VARIETIES

6. 'SONG OF INDIA' PLEOMELE (*D. reflexa*) has variegated leaves and needs regular pruning to maintain a pleasing shape.

7. 'FLORIDA BEAUTY' (*D. surculosa*), or gold dust dracaena, has wiry stems different from most dracaenas. It reaches 2 feet tall.

Fig
(*Ficus* spp.)

This diverse genus of more than 800 species is native to tropical and subtropical regions worldwide. The group includes trees, shrubs, and climbing or creeping vines. Most of the stately trees, from the ever-popular rubber tree (*F. elastica*) to edible fig (*F. carica*) can be kept 3 to 8 feet tall indoors. Most shrubby varieties grow up to 4 feet tall. Vines can be trained up trellises, used as groundcover, or grown as hanging plants.

Best site

Figs are among the most useful evergreen plants for leaf interest and tropical effect. Display a plant where you will get the best effect from its presence. Light requirements vary by variety, but most figs prefer medium light in spring and summer and bright light in fall and winter. Exceptions include the groundcover oakleaf fig (*F. montana*) and creeping fig (*F. pumila*); provide them with medium to low light. Figs are notorious for dropping their leaves if they don't receive enough light. They also react this way to sudden changes in the environment, cold drafts, and low winter temperatures, as well as too much water or fertilizer. If you're worried about your plant dropping its leaves, don't move it outside for the summer. Figs need consistent temperatures between 60°F and 80°F. Most varieties are content with average humidity (30–60 percent). High humidity is essential for vining figs.

Growing

Plant figs in compost-rich, well-draining potting mix. Small plants do best if repotted annually in a slightly larger pot. Larger plants can be kept in the same pot by changing the soil and pruning the roots. In general, figs need evenly damp soil during the growing season. Allow the soil to dry slightly between soakings during winter. Vining types require more frequent waterings. Fertilize a fig monthly throughout the growing season, using an all-purpose plant food. Most figs benefit from occasional pruning to maintain

TREELIKE VARIETIES

1. WEEPING FIG (*F. benjamina*) varies from small tabletop-size varieties to large floor plants. 'Starlight', pictured here, has green-and-white variegated leaves and arching branches.

2. 'ROBUSTA' RUBBER TREE (*F. elastica*) has extra-large burgundy leaves up to 18 inches long.

3. MISTLETOE FIG (*F. deltoidea*) grows to 15 feet tall. It has spoon-shape leaves with reddish undersides and tiny fruits.

4. 'BREEZE VARIEGATED' WEEPING FIG (*F. benjamina*) is noted for its resistance to leaf drop. This variety features small, glossy leaves edged in white.

5. 'AMSTEL KING' (*F. maclellandii*) impresses with its glossy, pinkish-reddish foliage on an upright, treelike plant form.

their form or to keep them smaller and within bounds. In general, propagate plants by stem cuttings, providing rooting hormone powder and humidity, or by air layering. It is natural for older shrub- or tree-type figs to shed their lower leaves gradually. Cut the bare stem of a rubber plant back to its base, and propagate the leafy tip to make a new plant. If a fiddleleaf fig (*F. lyrata*) loses its lower leaves, propagate stem cuttings or air-layer the plant. Prune wayward branches and trim stem ends to shape a weeping fig (*F. benjamina*), narrowleaf fig (*F. maclellandii*), or Indian laurel fig (*F. microcarpa*). These plants can be cut back by one third each year to keep them in good form.

Special notes

Wear gloves when cutting figs—the sap irritates the skin and eyes and can cause allergies. It is also slightly toxic, so keep plants away from children and pets. Scale insects are a common problem for figs. Mealybugs and spider mites may also become problems. Avoid using leaf polishes. Instead, dust leaves with a damp cloth. When bringing home a new plant, minimize the shock of a change to lower light conditions by draping the plant with a sheet of lightweight clear plastic for a few weeks. This raises the humidity until the plant adapts to the new setting.

8

6

7

VINING VARIETIES

6. VARIEGATED CREEPING FIG (*F. pumila* 'Variegata') has white leaf edges that contrast with the green of its foliage. Pair it with a Norfolk Island pine or grow it in a terrarium.

7. VARIEGATED TRAILING FIG (*F. radicans*) is a small, trailing vine with white-splashed 2-inch-long leaves. It needs more frequent watering than most other figs.

8. CREEPING FIG (*F. pumila*) is a woody, evergreen vine with 1- to 1½-inch-long leathery leaves, perfect as a groundcover or for covering a topiary form.

Flamingo Flower
(Anthurium andreanum)

Hundreds of species and hybrids of flamingo flower include both flowering and foliage plants. They bring tropical flair to the indoors. The best-known blossoming types produce a floral spike (spadix) and a colorful heart-shape leaflike bract (spathe).

Best site

Flamingo flower blooms can last for months, and many species are repeat bloomers when grown in bright light. Provide medium to bright light for flowering plants; foliage plants adapt to lower light. Hybrid flowering types have been developed for the average conditions of most homes and they are not difficult to maintain even in low light. Average to high humidity (up to 75 percent) is necessary for most flamingo flowers to survive indoors; some varieties require very high humidity (80 percent) and will not bloom well without it. Thin-leaf varieties need high humidity; leathery-leaf ones can cope with drier air. Research a variety to determine its needs. Maintain temperatures of at least 60°F at night and 70°F during the day. Keep the plants away from cold drafts and high-traffic areas—the exotic leaves and flowers are easily damaged.

Growing

The largest group of anthuriums are epiphytes with aerial roots. They need a coarse, fast-draining medium such as bark chips. The roots can then be covered with sphagnum moss. For other types, use a mixture of half potting mix and half small bark chips to make a well-draining medium. Repot the plant every other year or when the stems crowd the pot. Fertilize in summer only, using food formulated for flowering plants. Water the potting medium thoroughly, then let it dry slightly before the next watering.

Special notes

Propagate plants by division. Give your flamingo flower an occasional shower to prevent spider mites. Prune only to remove old flowers and leaves. Excess mineral salts cause leaf tips to brown and stems to die back. When repotting, place the crown high in the pot to help water drain easily away from the plant and to prevent root rot.

FLAMINGO FLOWER VARIETIES

1. 'SARA' (*A. andreanum*) has pale salmon-pink flowers four times the size of most varieties. It was first developed for the cut-flower industry.

2. 'RED ROCKET' (*A. andreanum*) is graced with dark red flowers and a white spadix. This blooming plant has good disease resistance.

3. 'WHITE HEART' (*A. andreanum*) combines large flower size with disease resistance. The white spathe contrasts with a pink spadix.

4. 'KERRICH RUBY' (*A. andreanum*) is known for its bright red spathe and spadix. Watch for this and other varieties in stores around Valentine's Day, Easter, and Mother's Day.

Flowering Maple

(*Abutilon* hybrids)

This adaptable plant grows easily and vigorously, up to 3 feet per year. Varieties have varying forms, from upright plants (shrubs or trees) that reach 4 to 5 feet to trailing, wiry-stem ones that star in hanging baskets. Most have maplelike leaves and showy cupped flowers in assorted colors.

Best site

Blooming occurs much of the year when light is medium to bright, with flowering often most intense during spring and summer. The plants stretch and stop blooming in low-light situations. Set flowering maple outdoors for summer, situated where it is shaded from midday and afternoon sun. The plants fare best in average temperatures (60°F–75°F) and no less than 50°F at night; they grow well in a cooler winter setting. Average to high humidity is ideal. Give the plants plenty of room to grow with good air circulation.

Growing

Flowering maple blooms best when somewhat pot-bound. The roots fill a pot quickly, so a plant may need repotting annually if it is ideally situated. Repot in winter or early spring to avoid disrupting the bloom cycle. Keep the potting mix evenly moist during the flowering season. Reduce water in winter, allowing the top 2 inches of potting mix to dry between waterings. Feed flowering maple monthly from spring through summer with a plant food formulated for blooming plants.

Special notes

Propagate flowering maple from stem cuttings or seeds. The plants grow quickly from seeds and this is a good way to get the cultivar you want. Remove spent flowers and yellowed leaves regularly. If you have a shrub- or tree-type plant, cut it back to half its size in fall to ensure a bushy, full plant the following year. Otherwise, cut stem tips back slightly in winter to shape the plant and encourage it to flower abundantly. Whiteflies and mealybugs are the most common pests to plague flowering maple. If your flowering maple doesn't bloom, set it in a place where it receives bright light for at least five hours each day.

YOU SHOULD KNOW

Grow cultivars of flowering maple with white variegation as foliage plants—they rarely bloom indoors.

FLOWERING MAPLE VARIETIES

1. VARIEGATED ABUTILON (*A. pictum*) is a colorful plant that has light orange bell-shape flowers and spring-green leaves heavily variegated with yellow.

2. 'MARILYN'S CHOICE' is a large, 4-foot-tall shrub that displays 2-inch-long, bell-shape blooms with red calyxes (blouses) on top of yellow-orange petals (petticoats).

3. 'NABOB' is noted for its nodding, deep crimson blooms, which are part of the charm of this 6-foot-tall hybrid shrub.

4. 'LUCKY LANTERN WHITE' is part of the Lucky Lantern Series; a dwarf with compact, bushy 12-inch-tall form and white blooms.

Grape Ivy
(*Cissus* spp.)

This fast-growing relative of grape shares the same habit of climbing via tendrils. Often grown in a hanging basket or trained up a trellis, grape ivy (*C. rhombifolia*) has long been a favorite indoor plant because it flourishes in a wide range of conditions. The dark green leaves add to the plant's lush appearance.

Best site

Place grape ivy in bright light during the winter and medium (indirect, bright) light the rest of the year. It also tolerates low light for extended periods, but pinch the vine ends frequently to keep the plant lush. Average humidity and temperatures of 60°F to 80°F are best, although grape ivy also adapts to cooler conditions (no less than 50°F). Ensure good air circulation to prevent powdery mildew, a fungal disease, and white flies, a common pest.

Growing

Grape ivy has a minimal root system and seldom needs repotting. When the plant cannot take up water, it's time to repot. Let the potting mix dry a little between waterings. The leaves will shrivel and die if the soil is too dry or too wet. Provide an all-purpose fertilizer during the growing season. Start new plants by making stem cuttings anywhere along the stems. Remove leaves from the bottom of the cutting and root in water or a soilless mix. Or propagate by layering. To rejuvenate a lanky plant, cut back about one third of the vines to 6 inches in late winter or early spring. One year later, cut back another one third of the plant's older vines; repeat the following year to complete the process. Practice this pruning regimen annually, cutting off no more than one third of the older vines to maintain a dense plant.

Special notes

Grape ivy also makes an attractive groundcover for a large potted plant, such as a Norfolk Island pine or weeping fig. Show off grape ivy by setting the potted plant on a pedestal, or training the vines onto a topiary frame and tying them in place with short lengths of soft twine. Shower the plant monthly to ward off spider mites. Keep an eye out for mealybugs, another common pest.

GRAPE IVY VARIETIES

1. GRAPE IVY (*C. rhombifolia*) grows as a vine well-suited to hanging baskets or as a specimen plant set on a pedestal.

2. 'ELLEN DANICA' OAKLEAF IVY (*C. rhombifolia*) has deeply cut leaves resembling those of an oak tree. It has a compact form.

3. REX BEGONIA VINE (*C. discolor*) has silver-patterned leaves with red undersides comparable to a fancy-leaf begonia. It needs constant warmth and high humidity.

Hibiscus
(Hibiscus rosa-sinensis)

Known as Chinese hibiscus and rose mallow, this woody shrub is a reliable and long-lasting flowering indoor plant. Hundreds of cultivars are available in different leaf sizes and shapes—some are variegated. The intensely colored, short-lived flowers have single or double petals. The plant blossoms for months.

Best site

Hibiscus craves at least a few hours of direct sun each day. The more bright light it gets, especially during the winter, the better it will bloom. In the summer, move the plant outdoors where it can bask in sun early in the day. Protect the plant from drying winds, cold drafts, and frost. It grows best in cool to average temperatures (55°F–70°F) and needs high humidity to keep the buds from dropping. Conditions that are too dry, too damp, or too cold cause leaf drop.

Growing

Water just enough to keep the soil from drying out during the winter. In summer, keep the soil evenly moist. Fertilize every other week throughout spring and summer, using plant food formulated for flowering plants. Repot your hibiscus when roots start to grow out of the bottom of the pot. Propagate by taking stem cuttings in summer. Establish three or four main branches and remove all others at the base. Cut back the main branches by one third in late winter to promote new growth, lushness, and blooms. The plant blooms on new wood, so avoid pruning off new growth or you will lose the flower buds. Hibiscus can be shrubby, with multiple stems, or a single-stem tree form. The plant grows from 2 to 10 feet tall and 2 to 8 feet wide, depending on the variety and the extent of pruning it receives.

Special notes

Hibiscus drops its leaves if conditions change, but the plant will regenerate leaves on old stems. Spider mites can become a problem in dry air, causing leaves and buds to yellow and drop off. A weekly shower helps keep the pests at bay. Whiteflies, aphids, mealybugs, and scale are also potential pests. Yellow leaves may indicate low nitrogen—adding compost to the potting mix will help fix the problem.

YOU _ SHOULD KNOW _
The flowers, buds, and young leaves of *Hibiscus rosa-sinensis* are edible. All add color and texture to salads.

HIBISCUS VARIETIES

1. LUAU LILIKOI YELLOW has vibrant sunny flowers on a compact 3-foot-tall shrub, spreading cheer throughout the summer.

2. 'TONGA WIND' bears bright red 6-inch-diameter flowers that contrast with the shrub's dark green foliage.

3. 'MANDARIN WIND', one of the Tradewind Series of hibiscus, is suited to a large pot and outdoor living as well as colorful indoor display.

4. 'TIKI TEMPTATION' has showy oversize flowers that bloom twice as long as traditional hibiscus varieties.

YOU
— **SHOULD KNOW** —
Give Japanese aralia a monthly shower to keep the large, shapely leaves looking their best and to control spider mite populations.

Japanese Aralia
(Fatsia japonica)

Popular since Victorian times, this large-leaf shrub reaches 6 feet or more. It is fast growing, durable, and tolerant of various environments. The naturally large plant can be kept smaller by regularly pruning long shoots that develop. Compact and white-variegated forms of Japanese aralia are available.

Best site

Japanese aralia is especially easy to grow in a cool, well-ventilated—even drafty—location indoors. It grows vigorously in medium to indirect, bright light, but also tolerates low-light conditions. Japanese aralia also grows outdoors in tropical locations (Zones 8 to 10). When thoughtfully combined with fine-texture shrubs and flowering plants, the form of Japanese aralia shines. The shrub is evergreen and thrives in part sun or shade. Plant it near a foundation or at the back of the border for a reliable stand of glossy green color.

Growing

Allow the soil to dry out between waterings for a plant grown in low light; keep the soil more evenly moist in medium or bright light. Water less during winter. Pinch off the stem tips to promote branching; prune to keep the plant 3 to 4 feet tall. Japanese aralia blooms in fall. The white blossoms are followed by black fruit.

Special notes

Leaves turn pale green and develop brown edges if the plant is underwatered. Overwatering rots the plant's roots. Propagate by stem cuttings.

JAPANESE ARALIA VARIETIES AND DETAILS

1. 'SPIDERS WEB' aralia offers palmate, or handlike, leaves etched in white.

2. WHITE FLOWERS look like little globes and appear in fall.

Lipstick Plant

(Aeschynanthus spp.)

This showy, trailing tropical plant from Southeast Asia produces scarlet flowers that unfurl like lipsticks rolled out of their tubes. The plant usually blooms most in fall then sporadically throughout the year. Lipstick plant has thick, glossy leaves and gently arching stems. A hanging basket or pedestal provides ideal displays.

Best site

Provide medium to bright light but not direct sun, average to warm temperatures (60°F–80°F), and high humidity (65 percent). Let the plant rest after blooming, cutting back on water, reducing temperature, and lowering humidity.

Growing

Allow the soil to dry slightly between waterings, but let it dry more during the winter rest period. If the leaves look wrinkled, then the plant isn't getting enough water. Feed every two weeks during bloom time with a formula for flowering plants. To raise the humidity of the air around lipstick plant, set the plant on a tray with pebbles and water. You can also use a humidifier. Regular misting helps maintain high humidity.

Special notes

Lipstick plant produces glossy 2- to 4-inch leaves; plants may shed leaves if the temperature is too cold. The leaves may also turn brown when humidity is too low. The vines reach 2 to 3 feet long. They can become leggy looking. Cut them off to 6 inches if they become scraggly and the vines will put on more robust growth and look more lush. Lipstick plant can be propagated by stem cuttings, in spring or summer. Repot plants in spring if plants seem excessively root bound. Move up only one pot size because plants like to be a bit root bound in order to bloom.

**YOU
— SHOULD KNOW —**
Grow new lipstick plants from stem cuttings taken in spring. Any variety makes a striking plant, even without flowers.

LIPSTICK PLANT DETAILS

1. FLOWERS are tubular and bright red. They look like a lipstick emerging from the tube. Most lipstick plants have red flowers, but some varieties have flowers that bloom in orange, yellow, and coral.

Lucky Bamboo
(*Dracaena sanderiana*)

Lucky bamboo is not a bamboo at all, but rather a dracaena with a pliable stem that is often woven into elaborate shapes. It grows quite well directly in water or gravel filled with water. This plant is reputed to bring good fortune (hence its name) and is often given as gifts.

Best site

Like most dracaenas, Lucky bamboo generally does best in medium light. Although most dracaenas can tolerate lower light, they may grow more slowly. Lucky bamboo grows best in average temperatures between 60°F and 80°F. In the winter, the plants adjust to cooler conditions (55°F). Most varieties need average humidity (35–65 percent). Dracaenas don't like cold drafts.

Growing

Lucky bamboo is frequently sold growing in water, although this plant can grow in wet sand or moist potting soil. If you grow lucky bamboo in water, make sure there is at least an inch of water touching the bottom of the stems. If grown in water, the plant will eventually require some fertilizer to maintain growth and leaf and stem color; Lucky Bamboos in need of nitrogen may turn yellow. Feed diluted liquid fertilizer every other month. Change the water frequently so bacteria doesn't grow, which could kill the plant.

Special notes

Lucky bamboo is sold in a variety of configurations, and grouped, braided and bound together with a band. To keep the original shape of the plant, remove new shoots with scissors or pruners. Lucky bamboo is easy to propagate in water.

LUCKY BAMBOO VARIETIES

1. VARIEGATA is a variegated lucky bamboo. Leaves are green and chartreuse. Stems are chartreuse. This plant is braided.

2. CURLED STEMS are just one of the many configurations of lucky bamboo. The more complex the design, the more expensive the plant.

Nerve Plant

(Fittonia albivenis)

The intricately veined leaves of this spreading 6-inch-tall plant inspire another name: mosaic plant. Varieties feature red, pink, or white veining on the oval leaves. All varieties, including a dwarf form, add color and structural interest to terrariums and dish gardens, where the conditions promote good health.

Best site

The plant does best in a warm and moist environment. Place it in medium, well-diffused light; direct sun burns the leaves. Temperatures above 65°F and humidity above 60 percent are ideal. For these reasons, it is a popular terrarium plant. It combines well in mixed baskets and groupings with other plants that require high humidity and soil moisture, such as small ferns and prayer plant. Place nerve plant at the edge of a container, where it will grow upright at first then begin to spill over the edge.

Growing

Nerve plant grows well when under a fluorescent light. It also adapts well to growing in an enclosed container such as a cloche. Use peaty potting soil that holds moisture well. Provide an even supply of moisture—not soaking wet—tapering off somewhat during the winter months and allowing the soil to dry a little between waterings. Fertilize monthly with all-purpose plant food.

Special notes

Pinch the stem ends regularly to keep nerve plant growing densely. When older plants lose their attractiveness, take stem tip cuttings and root them to make new plants. You can also propagate nerve plant by layering the stems. Sometimes the stems self-layer, forming roots where they touch the soil surface. Fungus gnats, mealybugs, and aphids are possible pests and root rot a potential disease of nerve plant.

YOU SHOULD KNOW

If you grow nerve plant in a wide pot, the low-growing stems will spread and root naturally, eventually filling the pot. This plant doesn't often need repotting.

NERVE PLANT VARIETIES

1. 'RED ANNE' foliage has wide pinkish-red veins set apart by olive green leaves, giving the plant an overall appearance of pinkish-red foliage.

2. VERSCHAFELTII also known as painted net leaf, adds its showy oval leaves with bright pink veins to any plant display.

3. ARGYRONEURA is a group of nerve plants that feature silver or white vein patterns on the olive green leaves.

4. 'MINI' has smaller leaves and a dwarf stature that make it an ideal plant for the confined spaces of a terrarium.

Palm
(Various genera)

Undemanding and adaptive, palms have long been favored as indoor plants. Many flourish indoors for decades, providing a lush, tropical backdrop. Their graceful leaves, called fronds, are either fan-shape or pinnate (feathery). Most palms grown indoors have a trunk, but others are clumping or stemless. Fronds range from 6 inches to many feet long. Your choice of plants depends on the shape and potential size that appeal to you.

Best site

Palms team up beautifully with other tropicals. Underplant a palm with colorful low-growing houseplants for a pretty effect. Although palms are among the world's most easily recognized plants, it is often mistakenly assumed that the plants require desert conditions with direct sun and dry heat. This family of plants that evolved millions of years ago includes thousands of species within different genera, so few generalities apply. However, most palms will grow best in medium or indirect, bright light. The few that thrive in low light include lady palm (*Rhapsis excelsa*), parlor palm (*Chamaedorea elegans*), and Kentia palm (*Howea forsteriana*). Typically, palms grow well in warm temperatures (75°F or higher) and high humidity (65 percent or higher), but many types can adapt to average temperatures and lower humidity. Very low humidity can cause frond tips to turn brown. Palms should not be situated in cold drafts. If you move a palm outdoors for the summer, keep it in a spot where it is shaded most of the day.

Growing

Usually slow-growing palms have varying needs. For best results, research the species of plant you want and address its particular needs accordingly. During spring and summer, water liberally to keep the potting mix evenly damp. Most palms will benefit from allowing the surface of the potting mix to dry out before watering during winter. The potting mix should drain well. Palms vary in their

TREELIKE VARIETIES

1. MAJESTY PALM (*Ravenea rivularis*) needs plenty of water, light, and fertilizer. It grows along rivers in nature. Spider mites can become problematic if humidity and moisture are lacking.

2. BAMBOO PALM (*Chamaedorea seifrizii, C. erumpens*), or reed palm, has a thick canopy. It likes uniformly damp but not wet soil.

3. CHINESE FAN PALM (*Livistona chinensis*) often grows as wide as it is tall. Give it bright, indirect light.

4. LADY PALM (*Rhapis excelsa*) is a slow grower that eventually may reach 10 feet tall. With its fan-shape leaves and multiple stems, it resembles bamboo.

5. KENTIA PALM (*Howea forsteriana*), or sentry palm, carries more fronds if kept in medium, indirect light.

needs for repotting. When the plant's roots fill the pot, move it into a somewhat larger pot with fresh potting mix. If a year or two goes by and the plant does not need repotting, replace the top 2 or 3 inches of potting mix with fresh mix. Palms do well in most potting mixes, but they'll grow better if you amend the usual mix with small pine bark and coarse sand or use a specialized palm potting mix. Fertilize palms monthly during the summer, using a plant food formulated for houseplants. Palms thrive and develop a rich green color when spending the summer outdoors in a shaded spot. Shelter your palm from wind to prevent damage to the fronds.

Special notes
Allow plenty of room for palms. They can be damaged easily. Snip off any damaged or dead fronds, but never cut off the top of a plant. Unlike most plants, palms produce new growth only from the tip, so removing its growing point kills the stem—and the whole plant. Some palms draw nutrients from old fronds (yellow or brown), so remove them judiciously and avoid taking off too many fronds, which will weaken the plant. Palms can be grown from seed, but most grow so slowly it makes sense to buy a palm the size you need. Shower the plant regularly, especially when the air is dry, to help prevent spider mites. Whiteflies, mealybugs, and scale may also present problems. Palms are sensitive to mineral salt accumulation in soil, from the local water supply or fertilizer. Use rainwater to hydrate a palm and help prevent this problem.

SHRUBBY VARIETIES

6. PYGMY DATE PALM (*Phoenix roebelinii*) is a slow grower that reaches 3 or 4 feet indoors and tolerates less than ideal conditions. It likes warmth and bright light.

7. PARLOR PALM (*Chamaedorea elegans*) prefers warm temperatures, but is not as fussy about high light and humidity as are many other palms.

8. ARECA PALM (*Dypsis lutescens*) produces yellow to yellow-orange lower stems. Give it bright light, high humidity, and consistent warmth.

Peace Lily
(Spathiphyllum wallisii)

The peace lily gets its common name from its distinctive bloom—a bract—that resembles a white flag. The plant has become popular for its glossy arching foliage and adaptable nature. Dozens of cultivars exist in varied sizes: dwarf (1 foot tall), medium (2 feet), and large (3 feet).

Best site

Peace lily is commonly used in commercial settings because it tolerates most indoor conditions. The plant grows best in medium to indirect or filtered, bright light. Light that is too intense yellows the leaves. Peace lily tolerates lower light but usually needs bright light to bloom. As long as the plant is kept out of cold drafts, it can cope with a range of temperatures (60°F–80°F). Placing the plant on a chilly windowsill (below 55°F) can damage it. The higher the humidity, the better, although peace lily is forgiving about this too.

Growing

If you grow peace lily in bright light, keep the soil damp. The lower the light, the drier the soil should be—especially in winter. The soil should not dry out completely, allowing the plant to wilt. If water is too scarce, the leaf tips may turn brown. Fertilize monthly in summer, using an all-purpose plant food. Overfertilizing will prevent flowering and cause brown spots on the leaves. Pot a peace lily in good-quality potting mix and repot rarely, when the roots have filled the pot. This is also a good time to divide the plant and pot the divisions in fresh soil.

Special notes

The plant typically blooms in spring and sometimes in fall. But many varieties, especially newer ones, have been bred to produce flowers more frequently. They may bloom sporadically throughout the year. Dust or shower the plant monthly. Most plants begin flowering only when they are more than a year old. Snip off yellow leaves and spent flowers. Insect pests are rare, but keep an eye out for potential problems, including mealybugs, spider mites, and scale. Overwatering and setting the plant too low in the pot can cause a peace lily's crown (raised center) to rot.

PEACE LILY VARIETIES

1. **'JETTYSTAR'** is a new cultivar with glossy, smooth, broad green leaves, clear white blooms, and compact growth suited for containers 8 to 10 inches in diameter.

2. **'DOMINO'** variegated peace lily is a compact plant that works well on a tabletop, where its white streaks can be appreciated up close. It grows 18 to 24 inches tall.

3. **'SENSATION'** is a massive cultivar that grows to 5 feet tall and 4 feet wide. The tough leaves have a ribbed texture.

Peacock Plant

(Calathea spp.)

This genus includes hundreds of beautiful foliage plants—some are challenging to grow indoors. Once you've had success with one, their appeal is strong. Peacock plants have ornately patterned leaves, marbled with pink, red, silver, or white. They mix well with ferns and other plants that thrive in humid conditions.

Best site

Peacock plant can be difficult to maintain without high humidity. Try grouping one—or several—with ferns in a bathroom. Use a humidifier to create ideal conditions (60 percent or higher humidity) and discover how other plants will benefit too. Or grow peacock plant in a tabletop conservatory or large terrarium. The plant grows best in medium light or diffused, bright light, but it can tolerate lower light when supplemented with artificial light. The coloring and patterning of peacock plant are most appreciable when lit from behind; the markings can fade in too-bright light. Many varieties have lacy, papery foliage or leaves with pale marbling set off by dark veins—reminiscent of stained glass. Provide warm temperatures (70°F–85°F); keep the plant out of cold drafts.

Growing

Keep the potting mix evenly damp from spring through fall. Use warm or room-temperature water. Fertilize monthly in summer, using an all-purpose plant food. Growth slows as winter comes, preparing the plant for a rest period. Continue to keep the humidity high, but reduce watering. Remove any dead leaves. Pinch off ragged leaves in late winter. Leaves will be replaced by new growth in spring. Divide the root ball every two or three years in spring or early summer and pot the divisions in well-draining potting mix to create new plants, if you wish. Shower the plant occasionally to remove dust; do not use leaf polish. Peacock plant grows from 1 to 2 feet tall, depending on the variety.

Special notes

Peacock plant may produce insignificant flowers, but dramatic foliage is the plant's greatest asset. Possible pests include spider mites, mealybugs, and aphids. Excess mineral salts cause leaf tip browning and stem dieback.

YOU — SHOULD KNOW —
Check a peacock plant's roots each spring. If they have filled the pot, it's time to give the plant a more spacious container with fresh potting mix.

PEACOCK PLANT VARIETIES

1. 'HOLIDAY' (*C. roseopicta*) is a spectacular plant with white-patterned green leaves with purplish undersides and pinkish-red flowers if grown in bright light.

2. 'ROYAL STANDARD' is a new hybrid variety with silvery leaves surrounded by a deep green margin. In bright light it may develop pink flowers.

3. CONCINNA PRAYER PLANT (*C. concinna*) has showy leaves that display a pattern of light and dark green. Small white flowers may appear at the base of the plant.

4. LANCELEAF CALATHEA (*C. lancifolia*), native to Brazil, grows to 30 inches tall with lance-shape leaves that have bold markings, purple undersides, and wavy edges.

Peperomia
(Peperomia spp.)

This extremely varied genus offers an amazing array of plants with diverse leaf forms, colors, and growth habits. Some peperomias grow upright, others trail, but most are compact and clumping. Some have flat, glistening leaves; others appear corrugated. Many have off-white flower spikes that resemble rat tails. These interesting plants display so much variety, an entire indoor garden could be created using peperomias alone.

Best site

Situate peperomias in medium or indirect, bright light. They usually grow well in a north- or east-facing window, or in a west window with a sheer curtain. Direct sunlight is too intense, but many varieties adapt to low-light conditions. They're ideal for a desktop or a dish garden. Variegated peperomias are less colorful in low light. As natives of the tropics and subtropics, peperomias fare best in temperatures between 60°F and 75°F, and should not be exposed to cold drafts or temperatures below 50°F. If leaf tips or edges turn brown, check the surroundings and move the plant away from a chilly windowsill, air-conditioning vent, or otherwise drafty spot.

Growing

The genus Peperomia includes more than 1,000 species, although most of them are not widely grown. Nonetheless, this is a fun group to collect—see how many different varieties you can find. They grow without special care and reward you with lasting beauty. Depending on the type you choose—upright, trailing, or clumping—the plants grow 6 to 12 inches tall. Some peperomias have waxy, fleshy, succulent leaves; the leaves of others are attractively marked.

SMOOTHLEAF VARIETIES

1. WATERMELON PEPEROMIA (*P. argyreia*) is a small species with silver-striped leaves that grows to 6 or 8 inches.

2. VARIEGATED BABY RUBBER PLANT (*P. obtusifolia* 'Variegata') has gold-and-white variegation on its glossy leaves. Try it in a terrarium.

3. 'PEPPY' (*P. ferreyae*) has thick, lancelike leaves. It grows 8 inches tall and 12 inches wide.

4. TEARDROP PEPEROMIA (*P. orba*) is a compact and mounding plant that begins to trail as it matures. Its leaves are shaped like an elongated teardrop.

5. RED-EDGE PEPEROMIA (*P. clusiifolia* 'Tricolor') bears leaves that have broad white borders with a pink blush along the edge. Pinch the stem ends to encourage branching.

Most peperomias will flourish during the warm, humid weeks of summer. But these adaptable plants can also tolerate the dry atmosphere of a heated home or office. Peperomias benefit from occasional showers to keep their leaves pristine. What's more, the plants tolerate a bit of neglect, such as a missed watering, better than most plants. For best results, allow the soil surface to dry between waterings, then water the plant from the bottom, avoiding wetting the leaves and the center or crown of the plant. Do this by filling the pot's saucer with water and allowing the potting mix and plant roots to draw up the moisture. Peperomias are sensitive to overwatering, which causes root rot, especially if the plant grows in cool and moist conditions. Keep the soil barely damp from late fall until early spring. Peperomia has a minimal root system, so repotting is not often necessary. After several years, when a plant has become crowded in its container, move it into a slightly larger pot with new potting mix. Use a peat-based potting mix that drains well rather than heavy potting soil. Feed peperomia every other week in the summer only, using a half-strength solution of all-purpose plant food.

Special notes

If your peperomia elongates into a heightened, central crown, it's time to make divisions of the root ball or remove offsets for new plants. Trailing peperomias are also propagated by stem cuttings. Peperomia is impressively pest-free most of the time. Scale and mealybugs become problems once in a while. If the leaves of your peperomia wilt or become discolored, or if the stems rot, chances are the plant is being overwatered. It may also be affected by ringspot, a viral disease marked by brown rings and deformed leaves. In any case, remove the damaged plant parts and cut back on watering. Ensure good air circulation around peperomia to help prevent disease.

TEXTURED-FOLIAGE VARIETIES

6. 'EMERALD RIPPLE' (*P. caperata*) has heart-shape, dark green leaves with a wafflelike texture. It is a popular easy-to-grow, compact plant.

7. SILVERLEAF PEPEROMIA (*P. griseoargentea*) reaches 6 inches tall and has silvery green leaves with a slightly rippled texture.

8. BURGUNDY (*P. caperata*) is one of the most popular varieties. It comes from the rain forests of Brazil and grows well in a terrarium or Wardian case.

Philodendron
(*Philodendron* spp.)

It's no wonder these are among the most widely grown indoor plants, known for their tough nature and tolerance of low light. Philodendrons' adaptability to a variety of settings, combined with a long life expectancy, make them highly successful indoor plants. In the wild, they grow on the floor of South American rain forests as vines, shrubs, and trees. With a conducive indoor environment, they typically pose few problems.

Best site

Philodendrons are grouped according to growth habits: climbing, tree, and clump-forming. The leaves of each are glossy and attractive—but different.

The most commonly grown climbing varieties do best in humid conditions and where they can attach themselves to a sturdy support, such as a moss pole or bark slab. Climbers also make excellent candidates trailing from hanging planters. Climbing philodendrons include spadeleaf (*P. domesticum*), splitleaf (*P. pertusum*), redleaf (*P. erubescens*), and heartleaf (*P. hederaceum oxycardium*, syn. *P. scandens*).

Tree philodendrons, in particular the cutleaf or lacy tree philodendron (*P. bipinnatifidum*), have a distinct trunk and can become tall (6 to 8 feet). They usually produce aerial roots, which reach into the soil to support the plant, and are especially suited to offices or large rooms with high ceilings.

Clump-forming philodendrons form ground-hugging rosettes that are often wider than they are tall. They rarely need staking and don't usually produce aerial roots, but some hybrids have climbing genes in their backgrounds and will eventually produce stems that begin to clamber. Clump-forming plants are handsome standing on a floor or a table.

CLIMBING AND TRAILING VARIETIES

1. 'BRASIL' has glossy leaves splashed with yellow and light green markings. This trailing plant reaches 4 feet long.

2. HEARTLEAF PHILODENDRON (*P. hederaceum oxycardium*, *P. scandens*) can be trained to a support or allowed to form a mounding plant.

3. VELVETLEAF PHILODENDRON (*P. hederaceum hederaceum*) is a graceful plant with a velvety leaf texture and bronze hues in the new foliage.

4. FIDDLELEAF PHILODENDRON (*P. bipennifolium*) needs medium light and high humidity. It will climb if you let it.

5. 'PINK PRINCESS' redleaf philodendron (*P. erubescens*) has black to burgundy foliage splashed with white and bright pink.

Philodendrons prefer medium light, but some tolerate low light. The heartleaf climber grows best in low light; with insufficient light others can develop smaller leaves that form farther apart on a stem. Generally, philodendrons grow well in average temperatures (60°F–75°F) and humidity (65 percent). Keep the plants out of cold drafts.

Growing

Allow the soil to dry slightly between waterings, especially in low light. During the winter, let the plant rest: Water only enough to keep the potting mix from drying out completely. Rinse the leaves periodically to keep them clean and glossy. To encourage growth, feed the plant monthly in summer, using an all-purpose fertilizer. Otherwise, feed the plant only once a year. Repot a philodendron only when the roots have filled the pot and the plant can no longer take up water. Do not use overly large pots or planters; philodendrons benefit from crowding. Use an average, well-draining potting mix. Cut back vigorous plants to control their size as you like, and snip off occasional faded leaves. Give climbing varieties adequate support with a moss pole—kept damp—or let them trail from a suspended container, and cut back the stems regularly to keep the plant lush.

Special notes

Propagate a philodendron from stem cuttings in summer. Keep the cuttings warm to promote rooting. Climbing varieties propagate well by layering the lengthy stems in soil. If a climbing plant becomes leggy or bare at its base, prune the stems to rejuvenate the plant and encourage new growth. Clump-forming plants may also become lanky over time. Cut them back by at least one third and up to two thirds—they'll return to their dense form eventually. Philodendrons are not ordinarily bothered by pests or diseases. If conditions are too dry, spider mites may proliferate. Mealybugs may also appear. All parts of a philodendron are toxic, and the sap irritates skin, so keep plants out of the reach of children and wear gloves while working with them.

8

6

7

SHRUBBY VARIETIES

6. 'CONGO' has large spadelike medium green leaves with lighter midribs. It withstands moderate light conditions. 'Rojo Congo' is similar, but with burgundy foliage.

7. 'PRINCE OF ORANGE' is a hybrid with coppery-orange new foliage. Leaves turn green as they age, but retain a pinkish midrib.

8. 'GOLDEN XANADU' has deeply lobed yellow-green foliage that forms a mounding plant up to 18 inches tall and wide.

Piggyback Plant
(Tolmeia menziesii)

Piggyback plant has fuzzy pale green leaves that often support little plantlets piggyback-style. The arching stems of this mounding plant grow to 12 inches tall and can quickly fill a hanging planter with an attractive mass of soft foliage. The weight of developing plantlets causes outer leaves to drape over the container's edge which makes piggyback plants a good choice for hanging baskets.

Best site

Place this plant in medium to bright light, not direct sunlight. It tolerates low light. It thrives in cool to average temperatures (60°F–75°F) most of the year and 50°F to 65°F in winter. Piggyback plant likes average humidity (30–60 percent). Mist occasionally for best health; leaf edges turn brown when humidity is too low.

Growing

Let the soil dry out a bit when growing piggyback plant in medium light; keep soil more moist when growing plants in bright light. Allow the soil surface to dry out between waterings. Feed plants every month in summer. Plants grow 12 inches tall and 18 inches wide.

Special notes

Rinse off the foliage under a fine spray or use a soft brush to remove dust. Propagate new plants by snipping leaves with "piglets" and potting them in small pots. Excess mineral salts damage the roots and cause the plant to die back. There are cultivars of piggyback plant that display bright chartreuse or golden-variegated leaves. Piggyback plant can cause skin irritation to individuals with sensitive skin.

PIGGYBACK PLANT DETAILS

1. PIGGYBACK "PIGLET" A mini leaf grows out of a larger leaf, giving the plant its common name, piggyback plant.

2. EASY TO PROPAGATE Just snip off a leaf with a "piglet" and pot it up in moist soil.

Ponytail Palm
(Beaucarnea recurvata)

The popular plant known as ponytail palm or elephant's foot is not a true palm but an agave family member, more closely related to dracaena and yucca. It resembles a ponytail, with narrow, dark green leaves that often reach down the full length of the trunk. It grows to 12 feet tall indoors.

Best site
This native of the semideserts of Mexico is impressively adaptable indoors, thriving in medium or bright light, dry air, and cold or heat. Provide the plant with bright light and keep it relatively dry. Normal room temperature is good for it most of the year, but in winter keep it cooler (50°F–55°F).

Growing
If this plant is flawed in any way, it is extremely slow growing. Water thoroughly, then let the potting mix dry between waterings. The plant stores water in its bulbous base. An established plant can go several weeks or longer without watering, especially when grown in cool conditions. The plant rarely needs repotting or fertilizing. If the plant is overwatered, the leaves will turn yellow. Err on the side of dry with this plant because ponytail palm rots when overwatered.

Special notes
Use a damp cloth to clean foliage monthly. Watch for spider mites, scale, and mealybugs. The leaves on ponytail palm often become brown at the tips, eventually turning completely brown as they age. Trim off the brown tips as they develop. It's best to prune off just the brown sections because any green leaf tissue feeds the plant. However, if you must trim more, do so by cutting on an angle for a natural pointed appearance to the leaf tip. Ponytail palm is sometimes sold as *Nolina recurvata*.

YOU SHOULD KNOW
Ponytail palm naturally sheds its long, strappy leaves as it grows. Otherwise, snip off any spent foliage as it turns brown.

PONYTAIL PALM VARIETIES AND DETAILS

1. 'GOLD STAR' is a colorful selection that sports leaves streaked in chartreuse. It does best in high-light situations and grows 8 feet tall.

2. A BIG BULB BASE gives the ponytail palm its distinctive look. The engorged trunk is where the plant stores water. Because of this, it doesn't require a lot of water.

Polka Dot Plant
(Hypoestes phyllostachya)

Also called Freckle Face, this spotted, dotted houseplant features colorful leaves with white, pink, or red splotches on a dark green background. One of the easiest houseplants to care for, polka dot plant is striking by itself in a pot or grouped with other houseplants.

Best site
Set plants in an area that receives medium or filtered light. In lower light, the plant's variegation may be lost. Polka dot plants prefer average temperatures (60°F–75°F). High humidity (60–70 percent) is important, so mist plants often.

Growing
Polka dot plant prefers rich, well-drained soil. Plants start out in a rounded form and stretch out as they grow. To keep plants in a mounded, shrubby form, pinch back new growth. Remove dead leaves to keep plants looking attractive. Fertilize containers once a month.

Special notes
Polka dot plants can be moved outdoors and planted in summer containers. They are originally from Madagascar and are actually a shrub; when they mature outdoors, they have a woody stem and can reach 3 feet tall. While frequently used outdoors, they generally are best when they are small; they make an attractive edging plant in larger pots because of their small stature and colorful leaves. Polka dot plant is easy to grow from seed or cuttings. Seeds germinate in warm, moist soils where temperatures are 70°F to 75°F; they sprout in just 14 days. You can clip and root cuttings any time, but they are most successful in spring and summer.

POLKA DOT PLANT VARIETIES

1. 'CONFETTI WHITE' has green veins and etching around the leaves. Some types have a dark green background with smaller dots and splotches.

2. 'CONFETTI PINK' features a beautiful combo of pink and light green.

3. PINK varieties have splotched leaves that may look more pink than green.

Pothos
(*Epipremnum aureum*)

Native to Southeast Asia, pothos is a vining plant widely used in hanging baskets or trained on moss stakes. Its heart-shape leaves are shiny green and often marked with yellow, white, or silver variegation. Pothos withstands neglect, low light, and poor watering practices. It's a good plant for novice gardeners.

Best site

Provide medium or indirect, bright light for best results. Direct sun will scorch the leaves. In lower light, the plant's variegation may be lost. Pothos prefers average temperatures (60°F–75°F), but will tolerate cooler periods. It is not so forgiving of cold drafts. High humidity (60–70 percent) is important. The leaf tips will turn brown and shrivel if the humidity is too low. When you grow pothos on a bark slab or moss pole, water the slab or pole to help the aerial roots hold on and provide more ambient humidity. To train pothos onto a moss pole or bark slab, insert the pole next to the root ball and drape the vines on it. You may need to occasionally pinch and reattach vines, but once the aerial roots form, the plant will keep itself upright.

Growing

Pothos may grow, as it does in the wild, as a climber. Its aerial roots cling to all but the smoothest surfaces, even on interior walls. Allow the potting mix to dry after being thoroughly soaked. Yellow and fallen leaves and rotting stems are indicators of overwatering. Feed pothos only once or twice a year with an all-purpose plant food. Repot every two or three years, when the plant's roots have filled the pot. Use a peat-based potting mix amended with coarse sand and perlite. Pinch off the plant's tips frequently to keep it lush. Periodically prune pothos as needed to maintain a full, symmetrical plant.

Special notes

Propagate pothos from stem cuttings taken in spring or summer. Root the cuttings in potting mix or water. You can also layer the stems (vines) to make new plants. Use 5-inch-long stem ends with remaining leaves to propagate new plants.

YOU
— SHOULD KNOW —
An old pothos plant may become too large or leggy for its space. When this happens, prune the plant back severely, to within an inch or two of the base.

POTHOS VARIETIES

1. COMMON POTHOS has deep green heart-shape leaves with splashes of gold.

2. 'NEON' is a trailing plant with bright chartreuse leaves. Because this variety grows to 2 feet or longer, it's an ideal candidate for a hanging basket in bright light.

3. 'MARBLE QUEEN' leaves are heavily variegated with creamy white. This variety needs a bit more light than others or it will lose its variegation and revert to green.

4. 'PEARLS AND JADE' has gray-green leaves with white splotches and a slightly smaller size (1 to 2 inches long and wide) than most other cultivars. It trails to 2 feet long.

Prayer Plant
(*Maranta leuconeura*)

Here's a must-have for a foliage plant collection, due to its unusual coloration and habit of rolling up at night. Various cultivars boast colorful markings. The undersides of the leaves are usually maroon. Prayer plant grows about 12 inches tall and wide and is well displayed in a hanging basket.

Best site
Prayer plant grows best with medium light and average temperatures (60°F–75°F). It can handle less light, but in excessively bright light the leaves brown along the edges and become pale. Protect the plant from cold drafts and keep the humidity high.

Growing
Evenly damp soil is ideal. Overwatering causes root rot, especially in winter. Fertilize monthly during the growing season with all-purpose plant food. Repot only when the plant fills its pot. Use well-draining, compost-rich potting mix.

Special notes
Rinse the plant monthly to remove dust and spider mites. Most prayer plant varieties can be propagated by division or stem cuttings.

PRAYER PLANT VARIETIES AND DETAILS

1. GREEN PRAYER PLANT is sometimes called rabbit tracks for its purplish-brown leaf markings. Closely related red prayer plant (*Maranta leuconeura erythroneura*) has bright red leaf veins along with the markings.

2. ROLLED LEAVES are one of the prayer plant's unusual features; leaves fold up at night. Some varieties feature maroon leaf backs, which add a bit of drama to the plant when the leaves roll up.

Purple Passion Plant

(Gynura aurantiaca)

Also known as purple velvet plant, this is an old-fashioned favorite. The velvety leaves of purple passion plant are densely covered with fuzzy purple hairs. New leaves are especially colorful. The fast-growing plant starts out growing upright. As it matures, the stems trail and reach 20 to 30 inches long. Purple passion plant provides a showy option for a hanging planter.

Best site

Give the plant bright light during winter and medium light the rest of the year. Purple passion plant becomes leggy in low light. If this happens, pinch back stems to promote bushier growth and move plant to a spot with more light. The plant does well with average temperatures (55°F–65°F) and humidity (30–60 percent).

Growing

Evenly damp soil is best; soggy soil rots the plant's roots. Purple passion plant can do without fertilizing and repotting. A happy, healthy plant produces clusters of yellow-orange flowers. Pinch off the flower buds because the flowers have an unpleasant odor and a mature plant declines after flowering.

Special notes

Purple passion plant is very easy to propagate. Make new plants from stem cuttings, which root easily in water or soil.

YOU
— SHOULD KNOW —
Regularly pinch out the stem tips of purple passion plant to promote branching and compact, lush growth.

PURPLE PASSION PLANT DETAILS

1. GOOD MIXER Purple passion plant works well in mixed containers. It offers subtle color, velvety texture, and vining growth habit.

2. PINCH BACK top growth if you want to keep the plant more shrubby than vining.

Purple Waffle Plant

(Hemigraphis alternata 'Exotica')

This low-growing, spreading plant reaches 6 to 9 inches tall and has deeply puckered leaves. The colorful foliage has a metallic purple sheen and burgundy undersides. The plant occasionally produces clusters of tiny white flowers. Display it in a hanging basket, as a groundcover for a large plant, or in a terrarium.

Best site

Grow purple waffle plant in medium light. Average to high temperatures (60°F and higher) and medium to high humidity (30 percent or higher) are best. Keep the plant out of winds and cold drafts.

Growing

The soil should be evenly damp. If it is allowed to dry, the leaves will turn brown and crispy; overwatering will cause the plant to rot. Repot annually in compost-rich potting mix. Feed monthly during the growing season with half-strength all-purpose plant food.

Special notes

Regularly pinching off the stem tips promotes dense growth and a well-balanced appearance. Root new plants from stem cuttings or by layering the stems in soil.

PURPLE WAFFLE PLANT VARIETIES AND DETAILS

1. 'EXOTICA' is a compact plant featuring crinkled green leaves with purple backs.

2. MODERN LOOK Purple waffle plants' metallic leaves add contemporary chic to decor. This plant is also called metal leaf and metallic plant.

3. GOOD MIXER Use purple waffle plant to add a burst of color in mixed planters. Here it is combined with ferns, moss, and amaryllis.

Sago Palm
(Cycas revoluta)

Although it resembles a palm tree, sago palm is more closely related to modern evergreens. Its unique architectural shape is an asset to any decor. Over time, the slow-growing plant develops a rough, thick trunk. The arching fronds are shiny and extremely stiff. Sago palm is a long-lived plant that grows to 6 feet tall.

Best site
Situate a sago palm in medium light year-round. If you move it outdoors for the summer, let it have morning sun and afternoon shade. Average indoor temperatures and humidity are fine. Because the fronds are so stiff and spiky, position the plant indoors where it can't be easily brushed against.

Growing
Let the surface of the potting mix dry between waterings. Avoid allowing the soil to dry completely. This plant does not indicate when it is dry by wilting (the fronds are very stiff), but the leaves will turn yellow if overwatered. Feed only once in spring and once in summer using half-strength fertilizer. Repotting is seldom needed because sago palm grows so slowly.

Special notes
The plant is toxic to children and pets, so display the plant out of reach if you have either in your household. The stiff, spiny fronds are easily damaged; position the plant in an area where it has lots of space around it.

YOU
_ SHOULD KNOW _
Avoid using pesticides on a sago palm. Many of them damage it. If necessary, turn to nontoxic pest control solutions instead.

SAGO PALM DETAILS

1. NEW GROWTH comes from the middle of the plant. The new leaves stand up like spikes.

2. FLOWERS appears on mature plants (generally 15 to 20 years old). There are male and female plants; therefore, flowers are different. This is a male flower.

YOU SHOULD KNOW

If you want a schefflera but are concerned about space, dwarf umbrella plant is smaller in leaf size (7 inches in diameter) but grows to more than 6 feet tall.

Schefflera

(*Schefflera* spp.)

The easy-care tropical plant, known as umbrella tree grows upright indoors, producing glossy leaflets on slender stems resembling small umbrellas. Indoors it grows to 8 feet but rarely blooms. As one of the most reliable, forgiving houseplants, schefflera can be a long-term evergreen resident.

Best site

Medium light produces the best growth, although schefflera can tolerate less light. It copes with high temperatures, but the plant will drop leaves if it gets too cold and grow spindly if the light level is too low. Average home temperatures (60°F–75°F) are ideal. Schefflera needs only average humidity (30–60 percent). Keep it out of drying winds and cold drafts.

Growing

Water regularly during the growing season to keep the soil mix evenly damp. Overwatering leads to leaf loss or root rot. In winter, water sparingly—just enough to prevent the soil from drying out completely. Fertilize monthly in summer; feeding more than that will spur an already vigorous plant and could translate into annual repotting. Adopt a schefflera only if you have room for one. To keep the plant smaller than a small tree, pot it in a container no larger than 8 inches in diameter. You may need to prune the plant's roots periodically. Replenish some soil annually. To keep schefflera looking its best, wipe leaves regularly with a dry, soft cloth. It is slow to propagate from stem cuttings; air-layering presents another option.

Special notes

Schefflera is often sold as several small plants in one pot—a good start for a lush appearance. The palmlike leaflets, small when young, can grow to 10 inches across. As the plant grows, you can shape it by pruning. A plant that becomes top-heavy should be pruned back to a more desirable shape. Remove wayward branches and faded leaves from time to time. Watch for mealybugs, scale, and spider mites. Give your schefflera a summer vacation outdoors, helping it adjust by exposing it gradually to the elements in a sheltered and shaded location.

SCHEFFLERA VARIETIES

1. **'AMATE SOLEIL' UMBRELLA PLANT** (*S. actinophylla*) is an improved variety with bright chartreuse foliage and has resistance to leaf spot and spider mites.

2. **UMBRELLA PLANT** (*S. actinophylla*) has treelike foliage consisting of 10-inch-long leaflets, attached to the stem like the ribs of an umbrella.

3. **'HAWAIIAN ELF' DWARF UMBRELLA PLANT** (*S. arboricola*) is a common indoor plant with dark green, shiny leaves. It is smaller in all dimensions than umbrella plant.

4. **FALSE ARALIA** (*S. elegantissima*, syn. *Dizygotheca elegantissima*) has much finer leaf texture than other scheffleras.

5. **'GALAXY' FALSE ARALIA** (*S. elegantissima*) is a bright color form of false aralia with creamy leaf edges.

Snake Plant

(Sansevieria spp.)

A strong accent plant, snake plant is prized for its ability to withstand low light, stifling heat, and even months of neglect. But it does better with proper care, a bright location, and regular watering. The tough, succulent leaves stretching to 5 feet tall grow from a rhizome.

Best site

Although snake plant can tolerate low light for extended periods, it performs best in medium or bright light. This succulent plant thrives in the average humidity (30–60 percent) and temperatures (60°F–75°F) of most homes, but it can cope with dry, hot conditions too. Snake plant even tolerates air-conditioning, but it will be happier outdoors in summer heat. During winter, a minimum temperature of 50°F is essential. Given bright enough light, the plant may produce a tall stalk of tubular, greenish flowers that exude a heavenly scent at night.

Growing

Let the soil dry between waterings. The plant rots easily if overwatered, especially in low-light situations. Water sparingly during winter. Grow snake plant in well-draining soil, amended with sand and pea gravel or perlite, and repot only when the plant fills the pot. Place more than one plant in a heavy shallow terra-cotta pot for an effective display. Fertilize only once a year, using an all-purpose plant food.

Special notes

Snake plant leaves break easily. If a leaf becomes damaged, selectively prune it out. Propagate snake plant by division, leaf cuttings, or leaf-section cuttings. Cut 3-inch-long leaf sections and stand a cut end in potting mix or vermiculite to root the cuttings. New plants started from leaf cuttings lose their variegation. Offsets or new plants also form at soil level and can be separated from the mother plant and transplanted. When dividing snake plant, cut a portion of rhizome with attached foliage, the surest way to reproduce variegated forms of the plant. Clean the leaves monthly to remove dust and keep them growing well. Remove any spent flowers. Pests rarely bother this plant.

YOU
_ SHOULD KNOW _
A wide variety of snake plant cultivars is available. Some have leaf markings with white, off-white, yellow, or chartreuse; others are green only. Some are compact; others are strongly upright.

SNAKE PLANT VARIETIES

1. SNAKE PLANT (*S. trifasciata*) has stiff sword-shape leaves that reach 4 feet and have gray horizontal markings resembling snakeskin.

2. VARIEGATED SNAKE PLANT (*S. trifasciata* 'Laurentii') is a popular variety with creamy yellow leaf margins. It grows 2 to 3 feet tall.

3. 'SILVER QUEEN' (*S. trifasciata*) is a cultivar with silvery gray foliage. In low-light conditions, the leaves become darker green.

4. 'HAHNII' (*S. trifasciata*) is a compact variety that produces rosettes with 4-inch-long leaves with horizontal stripes.

5. CYLINDER SNAKE PLANT (*S. cylindrica*) produces round, rigid leaves arching out from a central crown, eventually reaching 5 feet long and more than 1 inch in diameter.

1

Spider Plant
(Chlorophytum spp.)

One of the easiest and most popular indoor plants, spider plant grows quickly and adapts to cool or hot rooms, shade or sun. A hanging basket shows off the strappy, variegated foliage and long, wiry stems with attached dangling plantlets. Starry white flowers are a bonus of this South African native.

Best site
The plant will survive but not bloom or produce plantlets in very low light. It prefers indirect, bright light and may sunburn in direct sunlight. Spider plant grows vigorously when situated outdoors over the summer in a mostly shaded spot. Average temperatures (55°F–75°F) indoors suit the plant, with a winter minimum of 45°F. Average humidity suffices. If brown streaks appear on the leaves, the conditions are too cool and wet. Limp, pale leaves indicate too much heat and too little light in winter. The plant will improve in adjusted conditions.

Growing
Spider plant revels in regular showers. During the growing season, allow the soil to dry a bit, then soak it thoroughly. In winter, water sparingly. Fertilize spider plant every other month, using half-strength all-purpose plant food. This plant produces bulky, fleshy tuberous roots that are capable of storing water and enabling the plant to survive neglect. The roots can quickly fill a pot and make watering a challenge—water will run straight through the pot. Allowing the pot to soak up water from a deep saucer works well; pour off any excess after watering.

Special notes
Repot in spring, if needed. If excess mineral salts or fluoride causes the leaf tips to brown, repot the plant in fresh potting mix. Snip off the brown leaf tips. Starting new spider plants is easy. First, situate small pots full of potting mix next to the mother plant. Snuggle one or two plantlets into each pot. When the plantlets have rooted, cut the wiry stem to let the plantlets become fully independent. You can also divide a mature spider plant. Spring is an ideal time for propagating spider plant. A compost-and-peat blend works best for potting the plant; steer clear of a soil mix that contains perlite.

SPIDER PLANT VARIETIES

1. 'VARIEGATUM' SPIDER PLANT
(*C. comosum*) is among the easiest indoor plants to grow. This variegated type has ribbonlike green leaves with creamy white margins.

2. 'VITTATUM' SPIDER PLANT
(*C. comosum*) is a common cultivar with the reverse variegation of 'Variegatum'. It develops a central white stripe and green edges.

3. GREEN SPIDER PLANT (*C. comosum*), the plain species, has dark green, satiny leaves and is not as common as the variegated forms.

4. 'GREEN ORANGE' SPIDER PLANT
(*C. orchidastrum*, syn. *C. orchidantheroides*, *C. amaniense*) has broad foliage with orange petioles, which distinguishes it from its common relatives.

2

3   4

Strawberry Begonia
(*Saxifraga stolonifera*)

Neither a begonia nor a strawberry, this beautiful trailer produces scalloped silver-veined leaves that grow in a handsome mound with scads of plantlets. The leaves have red-tinged undersides. The delicate plant makes a pretty display in a hanging planter or perched on a pedestal. It reaches 9 to 12 inches tall, plus 12 inches of trailing stems and plantlets.

Best site
Medium to bright light brings out the best coloring in the fuzzy leaves of strawberry begonia. Give the plant bright light in winter and medium light the rest of the year. Too much direct sun can fade leaves or turn them brown. Site plants in areas where they receive cool temperatures (to 40°F) at night. Average to high humidity (30–70 percent) is ideal.

Growing
Strawberry begonia grows best in evenly damp soil with monthly doses of fertilizer throughout spring and summer. If leaves become crispy or shriveled, plants aren't receiving enough humidity. Use a humidity tray or room humidifier to raise the humidity around the plant. During the winter, allow the soil to dry out more and allow the plant to rest.

Special notes
Repot annually (spring is the best time) in well-draining potting mix. Strawberry begonia almost propagates itself with its tiny plants hung on delicate pink stolons. Leave the runners attached and pin (using a paper clip) the plantlets into pots surrounding the mother plant. Give the plantlets moist soil on which to sit and they will root. They will quickly form roots. After 3 weeks, snip the runners and you have a new plant. Low light causes spindly growth. Dry air and hot temperatures cause leaf dieback.

YOU
_ SHOULD KNOW _
Don't mist this plant because tiny hairs on the leaves can trap the water and cause the leaves to rot.

STRAWBERRY BEGONIA VARIETIES

1. VARIEGATED varieties offer a lighter look for strawberry begonia. Create a textural mix by combining variegated and nonvariegated in a container.

YOU
— SHOULD KNOW —
To promote stromanthe's best form, keep in mind that higher temperatures require high humidity to prevent leaf browning.

Stromanthe
(Stromanthe sanguinea)

The bold coloring of this little-known plant gives it strong appeal for indoor gardens. A relative of prayer plant, stromanthe bears rich emerald green- and ivory-streaked foliage that folds up at night. The colorful leaves reveal intense red undersides. The stems rise in a tight clump that fan out, reaching 12 to 18 inches.

Best site

Situate stromanthe in medium light, such as an east window, or indirect, bright light near a south- or west-facing window. Do not place plants in direct sun or leaves will look burned. Get best results in average temperatures (60°F–70°F) and average to high humidity (30–70 percent).

Growing

This plant needs evenly damp soil throughout the growing season. Ease up on watering in winter, allowing the soil surface to dry between waterings. Rinse the plant monthly (or wipe off leaves with a damp cloth) to keep it healthy and dust-free. Use an all-purpose fertilizer monthly in summer. If you move the stromanthe outdoors for the summer, place it in a shaded location. Stromanthe can grow outdoors year round in Zones 9 and higher. Elsewhere, take plants indoors before the threat of frost. Sometimes gardeners in the north grow this plant as an annual.

Special notes

Because this plant requires high humidity, place it on a tray with pebbles and water or add a room humidifier. Overly dry soil invites mealybug infestation. Overwatering causes stems and leaves to collapse or become diseased. Divide the plant to propagate it.

STROMANTHE VARIETIES AND DETAILS

1. 'TRIOSTAR' is a beautiful tricolor variety with white, green, and pink coloring. The backs of the leaves are red.

2. GOOD MIXER Stromanthe is an ideal large plant in a combo with 'Red Anne' fittonia and 'Royal Hustler' ivy.

Swedish Ivy
(*Plectranthus* spp.)

This fast-growing trailing plant is a member of the mint family—not an ivy. Stems drape well from a hanging basket and reach several feet long if unpinched. Swedish ivy produces a scent when the leaves are bruised or stems are cut. It develops spikes of tiny white or blue flowers.

Best site
Place Swedish ivy in bright light during the winter and medium light the rest of the year. It thrives outdoors during the growing season and will do best in a location with afternoon shade. Indoors, it prefers average temperatures (60°F–75°F) and humidity (30–60 percent).

The plant can handle cooler temperatures, but it must be protected from freezing. Many gardeners use cuttings from their indoor plants to provide outdoor plants for hanging baskets and other containers.

Growing
Swedish ivy grows well when the potting mix is kept evenly damp. Although it can survive drier conditions, long periods of dryness will cause the plant to produce smaller leaves. Fertilize monthly or less in summer, using a half-strength plant food. Give the plant plenty of room to grow in a spacious pot from the start and it will seldom need repotting. Use a compost-rich potting mix. By the time a plant outgrows its pot, it will likely be mature and no longer in top form—time to compost it. If you regularly start new plants from stem cuttings in soil or water, you'll always have attractive plants. They root and grow easily. Take cuttings or pinch off stem ends often to create a dense and compact plant.

Special notes
Most Swedish ivies grow upright at first, then arch over and eventually trail. Besides being ideal for hanging baskets, they also make a pretty display for a mantel, open shelf, window box, or mixed planting. Grow Swedish ivy as an understory to a large potted tree or train plant into a lush topiary. A sturdy plant that grows readily from cuttings, Swedish ivy gives even more pleasure when shared with friends and family. It is susceptible to whiteflies and mealybugs.

SWEDISH IVY VARIETIES

1. SWEDISH IVY (*P. verticillatus*) is a traditional indoor plant adapted to container gardening.

2. VARIEGATED PLECTRANTHUS (*P. coleoides* 'Variegata') pleases with its creamy white leaf edges. It is widely grown as an annual outdoors, adding trailing form to container gardens.

3. 'CERVEZA 'N LIME' is a hybrid plectranthus with fuzzy, succulent chartreuse foliage and a citrusy fragrance.

4. CUBAN OREGANO (*P. amboinicus*) also has succulent, velvety leaves with scalloped edges. This species has a sharp scent. It is valued as a culinary plant in some countries.

5. 'LIMELIGHT' (*P. oertendahlii*) is another easy-to-grow plant, characterized by its colorful foliage. The golden leaves are variegated with dark green and burgundy.

Swiss Cheese Plant
(*Monstera deliciosa*)

Easy to grow and durable, Swiss cheese plant is a dramatic addition to the indoor garden. The dark green, leathery, heart-shape leaves of a mature plant reach 3 feet across and have deep slashes and perforations. Aerial roots attach to a support and help the plant climb 10 to 15 feet. Its fruits are delicious.

Best site

Swiss cheese plant thrives best in medium to indirect, bright light. The plant produces smaller leaves without holes in lower light. Ensure average (60°F–75°F) warmth and humidity.

Growing

Let the top inch of potting mix dry between waterings. Water less in winter. Feed the plant monthly in summer. Repot infrequently in compost-enriched potting mix; top off the potting mix with an inch or two of compost annually. Give the plant room to grow and a strong support such as a moss pole.

Special notes

Clean the leaves regularly, using a damp cloth. Brown, brittle leaf edges indicate that humidity is too low; brown edges of yellowed leaves is symptomatic of overwatering. Propagate the plant by tip cuttings or air layering. All parts of the plant are toxic except the ripe fruit.

SWISS CHEESE PLANT VARIETIES AND DETAILS

1. 'CHEESECAKE' offers a variegated leaf; it's a showy collection of splotches and holes.

2. TRAINING UPWARD Swiss cheese plants can be trained onto a moss-filled post to make the plant grow more upright.

Ti Plant

(Cordyline fruticosa, syn. *C. terminalis)*

Also known as the good-luck plant, this sturdy tropical shrub grows upright with woody stems and reaches 3 to 6 feet tall indoors. The green or red leaves have pink or white variegations or margins, depending on the variety. Leaf color intensifies with age.

Best site
Indirect, bright light helps keep ti plant healthy. Direct sunlight scorches the leaves. High humidity is essential; use a humidifier during the winter and whenever the air is dry. Average temperatures (60°F–75°F) suffice.

Growing
Keep the soil evenly damp using rainwater, and allow the soil surface to dry between waterings. Apply all-purpose fertilizer monthly during the summer.

Special notes
Include several plants in a pot for a full appearance. Although ti plant is not usually bothered by pests, help prevent them by showering the plant monthly and removing yellow leaves. Brown leaf tips indicate low humidity or excess mineral salts or fluoride.

YOU
_ SHOULD KNOW _
Propagate ti plant by air layering or placing 2- to 3-inch-long pieces of stem cuttings in damp potting mix. Cover with plastic until the pieces sprout.

① ②

TI PLANT VARIETIES

1. 'KIWI' produces a colorful mix of green, white, and pink leaves. Use this colorful variety alone in a container or mixed with other houseplants.

2. 'EXCELSA' features dark leaves with white and pink splotches. Keep leaves clean to enjoy their glossy color.

YOU
— SHOULD KNOW —
A multitude of tradescantia cultivars boasts foliage striped with hues of green, purple, white, or silver.

Tradescantia
(Tradescantia spp.)

Few indoor plants are easier to grow or better suited for a hanging container. Tradescantia's foliage brightens an indoor garden almost as effectively as do flowers. This plant blooms, although the flowers are subtle. Tradescantia is mostly grown for its colorful leaves, trailing form, and ease of propagation.

Best site

Although tradescantia tolerates medium light, it grows well and maintains the best leaf color in indirect, bright light. If the light is too bright, the leaf color bleaches out; in low light, the leaf coloring can fade to green and the plant becomes leggy. Cool to average temperatures (55°F–75°F) help keep tradescantia more compact. Humidity of 30 percent or higher prevents the leaf edges from browning.

Growing

Tradescantia makes an attractive groundcover for larger houseplants, or the trailing element in a mixed planting with an upright plant and a midrange filler. You can perch the plant on a pedestal or shelf, as long as the stems can trail freely. Water enough to keep the soil evenly damp. Potting mix that is too wet or too dry causes the stems to deteriorate. Pot in an all-purpose, well-draining potting mix and repot only when the plant fills the pot with roots. During the growing season, fertilize monthly with half-strength plant food. Pinch the growing tips regularly to keep the plant compact and lush. Tradescantia does not age gracefully. Cut it back regularly and root the cuttings to provide a continuing source of new plants. When you take cuttings, tuck the cut stems into the same pot as the mother plant to make a fuller display, or root them in another pot.

Special notes

Tradescantia needs regular grooming to remove dead leaves and leggy stems and to encourage new, dense growth. Shower the plant to remove dust. Control mealybugs and spider mites; other pests are not common. Temperatures below 50°F will harm the plant; protect it from cold drafts. If the leaf tips or edges turn brown, check the humidity level and look for spider mites. Cut off stems with browned foliage.

TRADESCANTIA VARIETIES

1. WANDERING JEW (*T. zebrina*) is a traditional easy-care trailing plant bearing olive green foliage with silvery stripes and purple undersides.

2. PURPLE HEART (*T. pallida 'Purpurea'*, syn. *Setcreasea pallida*) is slow growing and requires less pinching than other species. It needs bright light to bring out the deep purple leaf color.

3. STRIPED INCH PLANT (*T. fluminensis*) is a fast-growing, trailing plant that forms a dense mass of shiny green-and-white foliage.

4. 'TRICOLOR' BOAT LILY (*T. spathacea*) is a striking white-, magenta-, and green-variegated form of boat lily with a finer texture.

5. BOAT LILY (*T. spathacea*) is a species that forms clumps of shiny green leaves with purple to maroon undersides.

Wax Plant

(Hoya spp.)

H. carnosa is the best-known member of this tropical plant genus. It is a climber with vining stems and waxy leaves. The growth habit, leaf shape, color, and size of each species differ. Some wax plants are compact; others are shrubby. Most have smooth, shiny leaves; others bear fuzzy foliage.

Best site

Wax plant rewards those who ensure its adequate light by producing fragrant blossoms. Train the plant on a trellis or grow it in a hanging pot. The vines can grow to 4 feet, and can be doubled on a support for a dense appearance. Place wax plant in indirect, bright light. Direct sun discourages flowering. The plant benefits from growing outdoors over the summer, as long as it is kept in partial shade or filtered light. Cool to average temperatures (45°F–75°F) will help the plant grow its best. Average humidity (30–60 percent) will do, but higher humidity prompts annual flowering. Situate the plant where you can enjoy the fragrance of the night-blooming flowers; avoid moving the plant once buds appear.

Growing

During the growing season, water wax plant only after the top ½ inch of soil has dried. In winter, when the plant is resting, water less but don't let the soil dry out completely. Repot the plant infrequently. It has a minimal root system and will bloom better when pot-bound. This plant does not withstand the stress of repotting well. Use an average soil that drains well and repot during mid- to late winter when the plant is resting. Fertilize monthly in summer with an all-purpose plant food.

Special notes

Overwatering causes leaves to drop. After the plant blooms, allow the faded flowers to fall off, but don't remove the flower stalk. Wax plant reblooms each year on the same stalk. Propagate by stem cuttings or layering the stems in soil. When cut, the stems of wax plant exude milky sap, which can attract insects. Keep an eye out for scale, mealybugs, and spider mites.

YOU SHOULD KNOW

Hoya is a genus that includes more than 200 species cultivars. The best conditions for growing each varies; most prefer bright light and relatively dry soil.

WAX PLANT VARIETIES

1. VARIEGATED WAX PLANT (*H. carnosa rubra*) is distinguished by its variegated leaves with green edges and creamy centers. This species produces pink flowers.

2. HINDU ROPE VINE (*H. carnosa* 'Crispa') has curled and twisted leaves packed tightly along the trailing stems.

3. 'ROYAL FLUSH' WAX PLANT (*H. lacunosa*) has elongated, flattened foliage heavily speckled with pale, silvery-green spots. New growth is purple, and it bears star-shape white flowers.

4. 'TRICOLOR' WAX PLANT (*H. carnosa*) has pinkish-purple stems and leaves variegated with pink, purple, and creamy white. Pink flowers appear in summer.

Zebra Plant
(*Aphelandra squarrosa*)

This showy plant is grown mostly as a dramatic foliage plant with shiny bold stripe leaves. In fall, zebra plant produces a bright yellow bract that lasts for months and shorter-lived tubular flowers in a similar hue. Compact, modern cultivars grow to 15 inches tall. The flowering spike adds several inches.

Best site
Zebra plant grows well in medium light from spring through fall and in bright light during the winter. It needs constant warmth (65°F or more) and high humidity (65 percent or more) to thrive. Cold, dry air scorches the leaves.

Growing
Keep the soil damp, but not soggy. Water less in winter when the plant rests. Feed only monthly during the summer, using half-strength fertilizer. When roots fill the pot, repot in spring.

Special notes
Zebra plant's leaves fall off when the soil is too dry or too wet or when the air is too dry or too cold. Adjust the environment and new leaves will develop to replace those lost to unfavorable conditions. The plant needs ideal growing conditions to rebloom, but the variegated foliage is attractive on its own.

ZEBRA PLANT DETAILS

1. FLOWERS of zebra plant are actually bracts. Flowers are usually yellow, but you can find zebra plants that bloom in orange or slightly red. The flowers naturally bloom in summer and stay in bloom for months.

2. LEAVES feature very defined stripes, which is where zebra plant gets its name.

Zeezee Plant
(Zamioculcus zamiifolia)

This plant is popular because it tolerates average household conditions and some neglect, yet keeps its attractive, glossy dark-green foliage. A distant relative of philodendron, it grows upright and forms a rosette of fleshy leafstalks. With time, it produces offsets and fills its pot with foliage.

Best site

Choose a location for the plant where it has room to grow and spread to its mature size: 3 feet tall and wide. It grows best in indirect, bright light, but can tolerate low light. Zeezee plant needs temperatures above 40°F and low humidity (30 percent).

Growing

Keep the soil damp in bright light; let it dry between waterings in lower light. Overwatering results in yellow leaves. Fertilize monthly in summer. Repot when roots fill the pot.

Special notes

Move zeezee plants outdoors to a shaded place for the summer. Shower the plant regularly or wipe the leaves with a damp cloth to remove dust. Propagate it from leaf cuttings, taking a full leaf rather than a leaflet for the cutting.

YOU SHOULD KNOW
Zeezee plant leaves are toxic. Take care to keep it away from pets and children.

ZEEZEE PLANT DETAILS

1. EASY CARE zeezee plants are exceptional choices for indoor gardeners who travel or are forgetful about watering. This plant is nearly indestructible.

2. GREEN GLOSSY LEAVES are thick and stiff. Keep plants free of dust to enjoy their natural lustre.

Index

A

Accents, architectural, 20
Acclimating plants, 39, 53
Aeonium, 103, 157
African milk tree (*Euphorbia trigona*), *26*
African violet (*Saintpaulia ionantha*), 23, 29, 33, 37, 50–51, 61, *136–137*
Agave (*Agave*), 157
Air-layering, *66, 67*, 207
Air plant (*Tillandsia*), 27, 76, 116, *117, 140*, 141
Air quality, improving with plants, 12–13, 172, 186
Alfalfa sprouts, *128*, 129
Algerian ivy (*Hedera canariensis*), 148
Almond, 101
Alocasia (*Alocasia*), 25, 62, **160**
Aloe (*Aloe vera*), 25, 157
Aluminum plant (*Pilea*), 51, **161**
Amaryllis, 10, 44, *198*
Arabian jasmine (*Jasminum sambac*), 151
Aralia (*Polyscias*), 16, *92*, **162**
Architectural plants, 20, 25
Areca palm (*Dypsis lutescens*), 13, 185
Arrowhead plant (*Syngonium podophyllum*), 19, 61, *109*, **163**
Arrowleaf fern, *118*
Artillery plant, *109*,161
Arugula, 130
Asparagus fern (*Asparagus densiflorus*), 62, 147
Avocado (*Persea americana*), 125
Azalea, 27, 37
Azuki bean sprouts, 129

B

Baby's tears (*Soleirolia solerolii*), 17, 108, **164**
Bacterial disease, 56–57
Balance, 27
Bamboo palm (*Chamaedorea erumpens*), 13, 184
Bark, 46

Barrel cactus (*Echinocactus grusonii*), 143
Basil, *122*, 123, 130, *131*
Basket gardens, 80–81
Bathroom, 23
Bay, 123
Bean sprouts, 129
Beets
  microgreens, 130, *131*
  sprouts, 129
Begonia (*Begonia*), 16, 19, 29, 33, 51
  angel wing, 139
  cane-stemmed, 139
  cuttings, 61
  growing indoors, 139
  iron cross, *138*, 139
  rex, *75*, *117*, 139
  rieger, 139
  terrariums, 108, 115, *117* , 139
  tuberous, 62, 139
Bell jars, *118*, 119
Biospheres, *114*, 115
Bird's nest fern (*Asplenium nidus*), 146–147
Bloodleaf (*Iresine herbstii*), **165**
Blushing bromeliad (*Neoregelia carolinae*), *56*, 141
Bonsai, 92–93
Boston fern (*Nephrolepsis exaltata*), 13, 54, 62, 147
Bottle gardens, *112*, 113
Boxwood, 89
Brake fern, *109*, 146–147
Branches, forcing, 100–101
Broccoli sprouts, 129
Bromeliads, 11, *56*, 68, 112, 115-116
  air plant, 27, 116, *117, 140*, 141
  lifestyles, 140–141
  misting, *50*
  offsets, 65, 140
  soil mix for, 46
  tank, 140–141
Brushing plants, 51
Buckwheat sprouts, 129
Bulbs, 10, *11*, 16, 78, 79, 80, 85
  forcing, *24*, 96–97, *98*, 99
  watering, 35
Burro's-tail (*Sedum morganianum*), 95, 157
Button fern, *112*

C

Cachepot, 17, 20, 41, 43, 80
Cactus
  caring for, 37, 143
  desert, *142*, 143
  dish gardens, 74, *75*
  handling, 143
  holiday, 10, 44, 95
  moving outdoors, 53
  offsets, 65
  soil mix for, 46–47, 143
  windowsill gardens, 72
Caddy, plant, 43
Caladium, *18*, 62
Calamondin, 126
Calla lily, 43
Care of plants, 31–57
  air circulation, 37
  disease prevention, 56–57
  fertilizing, *48*, 49
  grooming, 50–51
  humidity, 36–37
  light, 32–33
  mulching, 49
  pest control, 54–55
  potting and repotting, 44, *45*
  selecting healthy plants, 38–39
  symptoms of stress, 37
  temperature, 37
  transitioning plants, 39
  vacation care, *52*, 53
  watering, 34–35
Cast-iron plant (*Aspidistra elatior*), 16, 62, **166**
Centerpieces, *84*, 85–87, 133
Charcoal, 47, 74, 113
Cherry, 101
Chinese evergreen (*Aglaonema*), 12, 13, 17, *45*, 62, **167**
Chinese fan palm (*Livistona chinensis*), 184
Chirita (*Chirita*), 137
Chrysanthemum, 10
Cilantro, 123
Citrus, 126–127
Clivia (*Clivia miniata*), 44, **168**
Cloches, *118*, 119
Club moss, *112*,
Coffee plant (*Coffea arabica*), **169**
Coir (coconut husk), 46
Coleus (*Coleus*), 17, 51, 144, *145*

Color, 17
Compost, 46, 49
Conifers, dwarf, 76–77
Conservatories, tabletop, 108, *109*
Containers
  baskets, 41, 80–81
  bonsai, 92–93
  cachepot, 41
  choosing, 40–41
  creative, 82–83
  displaying pots, *42*, 43
  drainage, 40
  finding special, 83
  grouping plants, 19
  hanging planters, *94*, 95
  materials, 41
  miniature garden, 76–77
  seed-starting pots, *68*
  self-watering buckets, 53
  strawberry jar, 82–83
  style, 20
  water requirements, 34
  wooden planters, 41, *78*, 79
Corn plant (*Dracaena fragrans*), 13, 172
Creeping fig (*Ficus pumila*), 95, 174–175
Crocodile fern (*Microsorum musifolium 'Crocodyllus'*), 147
Crocus, 10, *11*, 86, 97
Croton (*Codiaeum variegatum pictum*), 13, 39, 67, **170**
Crystals, water-retentive, 46
Cuttings, 60–61
Cyclamen, *8*, 17, *21*, 27, 37, 85
Cypress, 89

D

Daffodils, *8*, 10, *11*, 80, 96–97
Date, 124
Decor, 15–29
  adding plants, 18
  balance, 27
  design basics, 16–17
  grouping plants, 19
  growing conditions, room-specific, 22–23
  plant size, 26–27
  previewing plants at nursery, 19
  siting plants, *24*, 25
  style, 20–21

Desktop garden, 18
Dieffenbachia (*Dieffenbachia*), 67, **171**
Dill, 123
Dining room, 23
Diseases, 56–57
Dish gardens, 74, *75*
Dividing plants, 62, *63*
Dracaena (*Dracaena*), 19, 25, 34, 39, 67, *82*, 108, *117*, **172–173**
Dusting plants, 51
Dwarf myrtle (*Myrtus communis*), *84, 85*
Dwarf papyrus (*Cyperus*), 74
Dwarf sweet flag (*Acorus*), 108
Dyckia (*Dyckia*), 140

E

Earth star (*Cryptanthus bivittatus*), 115, *118*, 140–141
Easter lily, 10
Eastern redbud, 100
Echeveria (Echeveria), 103, *156*, 157
Edible plants, 121–133
  citrus, 126–127
  exotic/tropical, 124–125
  herbs, *122*, 123
  microgreens, 130, *131*
  produce, 124–125
  sprouts, 128–129
  wheatgrass, *132*, 133
English ivy (*Hedera helix*), 13, 148
Entryway, 22–23
Environment
  adjusting plants to outdoors, 53
  air circulation, 37
  humidity, 36–37
  ideal, 28–29
  light conditions, 29, 32–33
  microclimates, 28–29, 37, 44
  room-specific, 22–23
  symptoms of stress, 37
  temperature, 37
Epiphytes, 46, *117*, 141, 152, 176
Eugenia (*Syzygium*), *88*
Euonymous topiaries, 76, 89
Evergreens, dwarf, 85, 87
Exotic plants, edible, 124–125

F

Fairy gardens, 76–77
False aralia (*Schefflera elegantissima*), *112, 117*, 162, 200
Felt, 43
Ferns, 34, 37, 43, 44, 47, 146–147
  dividing, 62
  seasonal centerpiece, 86
  soil mix for, 47
  terrariums, *8, 108, 109, 110, 112, 118*
Fertilizer, 49
Fertilizing, 46, *48*, 49
  bonsai, 93
  overfertilizing, 49
  seedlings, 69
  topiaries, 89
Fiber optic grass (*Isolepis*), 74
Fiddleleaf fig (*Ficus lyrata*), 67, 175
Fig (*Ficus*), 27, **108, 174–175**
  creeping, 95, 174–175
  large, 25, 26
  propagation, 67
  siting plants, 25
  weeping fig, 13, 25, 26, 174–175
Flamingo flower (*Anthurium andreanum*), 13, 19, 75, **176**
Flowering maple (*Abutilon*), **177**
Flowering plants, soil mix for, 46
Fluorescent lights, 33
Forcing branches, 100–101
Forcing bulbs, 96–97, *98*, 99
Forcing hyacinths, *98*, 99
Form, 16
Forsythia, *99*, 100–101
Fragrant plants, 24
  flowering vines, 151
  rosemary topiary, 89
  scented topiary ring, 91
Fritillaria, 80, 96
Fungal disease, 56–57
Fungus gnats, 55

G

Geranium, 19. *See also* Scented geranium
Germination, 69
Ghost plant (*Graptopetalum*), 157

Ginger (*Zingiber officinale*), 124–125
Glass, gardens under, 105–119
  biospheres, *114*, 115
  bottle/jar gardens, *112*, 113
  cloches, *118*, 119
  glass bubbles, 116, *117*
  moss gardens, *106*, 107
  tabletop conservatories, 108, *109*
  terrarium, *8*, 110–111
Globe terrariums, 116, *117*
Grape hyacinths, 8, 96
Grape ivy (*Cissus*), 95, **178**
Grooming plants, 50–51
Grouping plants, 19, 36, 143, 152
Growing conditions, room-specific, 22–23
Guava, 124
Guppy plant (*Nematanthus gregarius*), 115
Guzmania (*Guzmania*), *112*, 141

H

Habitats for plants, 28
Hanging planters, *94*, 95
Herbs, *24, 72, 122*, 123
  containers for, *11*
  environmental conditions for, *29*
  grouping, 19
  miniature garden, 76
Hibiscus (*Hibiscus rosa-sinensis*), 50, 54, **179**
Holiday plants, 10
Honeysuckle, 100
Horticultural oil, 55
Humidity, 36–37
Hyacinth, *8, 11*, 81, 96, *98*, 99
Hydrangea, 43
Hygrometer, 36

I

Impatiens, 108
Indian laurel fig (*Ficus microcarpa*), 174–175
Insecticidal soap, 55
Insectivorous plants, *118*
Insect pests, 54–55
Ivy (*Hedera*), *8, 24*, 95, 148–149
  English ivy, 13, 148
  topiary, 149

J

Jade plant (*Crassula ovata*), 103, 157
Japanese aralia (*Fatsia japonica*), **180**
Japanese holly fern (*Cyrtomium falcatum*), 146–147
Japanese serissa, 93
Jars, gardens in, *112*, 113
Jasmine (*Jasminum*), 95, *150*, 151

K

Kalanchoe (*Kalanchoe*), *27*, 43, 65, *72, 75*, 103, 157
Kale, *131*
Kentia palm (*Howea forsteriana*), 184
Kitchen, 23
Kiwi, 124
Kumquat, 126

L

Lady palm (*Rhapsis excelsa*), 13, 184
Lavender, 89, 123
Layering, *66*, 67
Leaf cuttings, 61
Leaf mold, 46, 146
Leaf spot, 57
Lemon butter fern (*Nephrolepis cordifolia*), 147
Lemon, 'Meyer,' 126
Lemon verbena, 123
Lentil sprouts, 129
Lettuce, 130
Lichen, *106*
Light
  intensity, quality, and duration, 33
  levels, 29, 32
  plants for bright, medium, and low, 29
  supplemental, 33
  symptoms of insufficient or too much, 32
  windowsill gardens, 73
Lilies, 10, 43
Lily-of-the-valley, *24*, 82–83
Lime, *124*, 126
Lipstick plant (*Aeschynanthus*), **181**
Living room, 22–23

Location of plants
  ideal environment, 28–29
  light conditions, 29
  microclimates, 28–29
  siting plants, **24**, 25
  specific rooms, plant-possibilities
     for, 22–23
Loquat, 124
Lucky bamboo (*Dracaena
  sanderiana*), *82*, 172–173, **182**

M

Madagascar dragon tree (*Dracaena
  marginata*), 13, 172–173
Madagascar jasmine (*Stephanotis
  floribunda*), 95, 151
Magnolia, 100
Maidenhair ferns (*Adiantum
  pedatum*), *28*, 43, *109*, *112*,
  146–147
Majesty palm (*Ravenea rivularis*), 184
Mango, 124–125
Mealybug, *54*, 55
Mexican butterwort, *118*
Mexican (key) lime, 126
Microclimates, 28–29, 37, 44
Microgreens, 130, *131*
Micronutrients, 49
Miniature gardens, 76–77
Misting, *50*
Mock orange (*Philadelphus*), 100
Moss gardens, *106*, 107
Moss pole, 90
Mother of pearl plant, 103
Moving plants outdoors, *52*, 53
Mulching, 49
Mung bean sprouts, 129
Myrtle (*Myrtus*), 76, 89
Myrtleleaf orange, 126

N

Narrowleaf fig (*Ficus maclellandii*),
  174–175
Nature, bringing indoors, 9
Nerve plant (*Fittonia albivenis*), 51,
  108, *110*, **183**
Nitrogen, 49
Norfolk Island pine, 19, 26

O

Oakleaf fig (*Ficus montana*), 174
Offsets, *64*, 65, 140
Orange, 126
Orchids, 19, 25, 26, 37, 90,
  152–153
  soil mix for, 46
  in Wardian case/terrarium, 108,
     109, *110*, *118*
Oregano, *122*

P

Pachyveria (*Pachyveria*), *156*
Palm, 13, 25, 34, 44, **184–185**
Pansy, *11*, *84*, 85
Papaya (*Carica papaya*), 124–125
Paperwhite narcissus, *84*, 85, 97
Papyrus (*Cyperus*), 108
Parlor palm (*Chamaedorea
  elegans*), 184–185
Parsley, *122*
Passionflower (*Passiflora*), 151
Passion fruit (*Passiflora edulis*),
  124, 125
Peace lily (*Spathiphyllum wallisii*),
  12, 13, 16, 37, *94*, **186**
Peacock plant (*Calathea*), **187**
Peanut cactus, 65
Peat moss, 46, 47
Pebble tray, 36
Pedestal cake plate, 107, *118*
Peperomia (*Peperomia*), 16, 51,
  61, 62, *117*, **188–189**
Perlite, 46
Pest control, 54–55
Petroleum jelly, 55
Philodendron (*Philodendron*), 13,
  25, 39, *52*, 67, 90, 95, **190–191**
Phosphorus, 49
Piggyback plant (*Tolmeia
  menziesii*), *64*, 65, **192**
Pinching back plants, 50–51
Pineapple (*Ananus comosus*), 125,
  140
Pitcher plant, *118*
Plantlets, *64*, 65
Plant stands, *42*, 43
Pleomele (*Dracaena reflexa*),
  172–173

Plum, 101
Poet's jasmine (*Jasminum
  grandiflorum*), 151
Poinsettia, 10
Poisonous plants, 13
Polka dot plant (*Hypoestes
  phyllostachya*), *69*, 117, **194**
Pollination, 127
Pollutants, indoor, 12-13
Pomegranate (*Punica granatum*),
  93, 125
Ponytail palm (*Beaucarnia
  recurvata*), **193**
Potassium, 49
Pot feet, 43
Pothos (*Epipremnum aureum*),
  13, 16, 19, *29*, 51, *52*, 62, 67,
  90, **195**
Pots. *See* Containers
Potting mix
  bonsai, 93
  soilless, 46–47, 113
  soil mixes, 46–47
  terrarium, 111, 113
  water requirements, 34
Powdery mildew, *37*, 56–57
Prayer plant (*Maranta leuconeura*),
  109, *187*,**196**
Prickly pear cactus (*Opuntia
  microdasys*), 143
Primrose, 16, 81, 86, 108
Produce, plants for, 124–125
Projects, 71–103
  basket gardens, 80–81
  bonsai, 92–93
  creative containers, 82–83
  dish gardens, 74, 75
  forcing branches, 100–101
  forcing bulbs, 96–97, *98*, 99
  hanging planters, *94*, 95
  miniature gardens, 76–77
  seasonal centerpieces, *84*, 85–87
  succulent wreaths, *102*, 103
  support systems, 90–91
  topiary towers, *88*, 89
  windowsill gardens, 72–73
  wooden planters, *78*, 79
Propagation, 59–69
  African violets, 137
  coleus, 144
  cuttings, 60–61
  dividing plants, 62, 63

layering, *66*, 67, 207
plantlets and offsets, *64*, 65, 140
from seeds, 68–69
Pruning, 50–51
  bonsai, 93
  topiaries, 89
  training to supports, 90
Purple passion plant (*Gynura
  aurantiaca*), 51, 61, 75, **197**
Purple waffle plant (*Hemigraphis
  alternata* 'Exotica'), **198**
Pussy willow (*Salix*), *78*, 79, 85,
  100–101
Pygmy date palm (*Phoenix
  roebelinii*), 185

R

Rabbit's foot fern (*Phlebodium
  aureum*), 43, 146–147
Radish, 129, *131*
Rattail cactus (*Aporocactus
  flagelliformis*), 143
Red amaranth, 130
Red clover sprouts, 129
Red flame ivy (*Hemigraphis
  alternata*), 198
Reindeer moss, *8*
Repotting, 44, *45*
  bonsai, 93
  large plant, 127
  orchids, 153
  for pest control, 55
Rex begonia (*Begonia rex*), 75,
  117, 139
Rhizomes, dividing, 62
Rhododendron, 100
Root cuttings, 61
Rooting hormone, 61
Rooting mix, 61
Rosemary, 123
  hanging planter, *95*
  miniature garden, 76
  topiary, 89
Roses, mini, 85, *113*
Rot, 57
Rubber tree (*Ficus elastica*), 67, 174

S

Sage, *95*
Sago palm (*Cycas revoluta*), **199**
Salmonella poisoning, 129
Sand, coarse, 47
Santolina (Santolina), 89
Saucers, 43
Scale, 55
Scented geranium (*Pelargonium*), 89, 91, *118*, 154–155
Schefflera (*Schefflera*), 13, 37, 83, 93, **200**
Scotch moss (*Sagina*), 77, 108
Seasonal centerpieces, *84*, 85–87
Seasonal plants, 10
Sedum (*Sedum*), *156*, 157
Seeds
    for edible gardens, 124–125
    microgreens, 130
    for sprouting, 128–129
    starting plants from, 68–69
    wheatgrass, 133
Selecting healthy plants, 38–39
Sempervivum (*Sempervivum*), 157
Sheet moss, 81
Shelving, 73
Siberian squill (*Scilla siberica*), 80
Silver vase plant (*Aechmea fasciata*), 141
Siting plants, *24*, 25
Size of plants, 26–27
Snake plant (*Sansevieria*), *12*, 13, 16, 17, 34, 62, **201**
Soaps, 55
Soil layering, 67
Soil mixes, 46–47
Sphagnum moss, 36, 103, *118*, 146, 152
Spider mites, 55
Spider plant (*Chlorophytum*), 13, 37, 44, 52, *64*, 65, 95, **202**
Spinach, 130
Sprouts, 128–129
Squill (*Scilla*), 80, 96, *156*
Staghorn fern (*Platycerium bifurcatum*), 95, 147
Staking plants, 90, *91*
Standard, coleus, 144, *145*
Stands for pots, *42*, 43
Starfruit (*Averrhoa carambola*), 125
Star of Bethlehem (*Ornithagalum umbellatum*), 17
Sticky trap, 55
Strawberry begonia (*Saxifraga stolonifera*), 65, 115, **203**
Strawberry jar, 82–83
Streptocarpus (*Streptocarpus*), *8*
Stress, symptoms of, 37
String of pearls (*Senecio rowleyanus*), 157
Stromanthe (*Stromanthe sanguinea*), **204**
Style, 20–21
Succulents, *8*, 34, 37, 61, 156–157
    containers for, *20*
    dish gardens, 74, *75*
    offsets, 65
    planting ideas, 157
    soil mix for, 46–47
    windowsill gardens, *72*
    wreaths, *102*, 103
Support systems, 90–91
Swedish ivy (*Plectranthus*), 17, 51, 61, **205**
Sweet potato (*Ipomoea batatas*), *124*, 125
Swiss cheese plant (Monstera deliciosa), 90, **206**

T

Tabletop conservatories, 108, *109*
Tabletop garden, *8*, 19, 25
    dish gardens, 74, 75
    miniature gardens, 77
    wheatgrass, 133
    wooden planters, *78*, 79
Tanglefoot, 55
Temperature, 37
Tepees, 90
Terrarium, *8*, 110–111
    bottle/jar gardens, *112*, 113
    condensation in, 111, 115
    glass bubbles, 116, *117*
    mini biospheres, *114*, 115
    planting, 111
    troubleshooting, 115
    watering, 113
Texture, 16
Thyme, 76, *122*

Ti plant (*Cordyline fruticosa*), 61, *66*, 67, **207**
Topiary, 27, 123
    ivy, 149
    miniature garden, 76
    scented ring, 91
    towers, *88*, 89
Toxic plants, 13, 163, 175, 182, 191, 199, 206, 211
Tradescantia (*Tradescantia*), 65, *94*, **208**
Trailing clubmoss (*Selaginella kraussiana*), 115
Training
    bonsai, 93
    to supports, 90–91
Transitioning plants, 39
Trays, 43
Trees, bonsai, 92–93
Trellis, 90, *91*
Tropical plants, edible, 125
Tubers, dividing, 62
Tulips, 19, 96, 101

U–V

Umbrella plant (*Schefflera actinophylla*), 200
Vacation care of plants, *52*, 53
Venus flytrap, *118*
Vermiculite, 46
Viola, *8*, *11*
Viral disease, 56–57
Vriesea, Red Flame, 25

W–Z

Wardian cases, 108, *109*
Water, types of, 35
Watering, 34–35
    hanging plants, 95
    method, 35
    terrariums, 113
    when to water, 34
    wick-watering system, 53
Water lettuce, *18*
Water meter, 35
Waterproofing a basket, 80

Wax begonia (*Begonia semperflorens-cultorum*), 139
Wax plant (*Hoya*), 44, 151, **209**
Weeping fig (*Ficus benjamina*), 13, 25, 26, 39, 174–175
Wheatgrass (*Triticum aestivum*), *24*, *132*, 133
Whiteflies, 55
Wick-watering system, 53
Windowsill gardens, 19, 25, 72–73, *122*, 123–124, 130
Winter dormancy, 93, 196
Winter (Chinese) jasmine (*Jasminum polyanthum*), *150*, 151
Wire, 90
Wooden planters, 41, *78*, 79
Workspaces, 23
Wreaths, succulent, *102*, 103
Year-round gardening, 10
Zebra plant (*Aphelandra squarrosa*), **210**
Zeezee plant (*Zamioculcus zamiifolia*), 62, **211**